AF412687

TEST METHODS

TEST METHODS

VOLUME 6

R. BYRON PIPES
ROBERT A. BLAKE, JR.
JOHN W. GILLESPIE, JR.
LEIF A. CARLSSON

Reviewing Editors

LEIF A. CARLSSON, Ph.D.   **JOHN W. GILLESPIE, JR., Ph.D.**

TECHNOMIC
PUBLISHING CO., INC.
LANCASTER · BASEL

Delaware Composites Design Encyclopedia—Volume 6
a **TECHNOMIC**® publication

Published in the Western Hemisphere by
Technomic Publishing Company, Inc.
851 New Holland Avenue
Box 3535
Lancaster, Pennsylvania 17604 U.S.A.

Distributed in the Rest of the World by
Technomic Publishing AG

Main entry under title:
  Delaware Composites Design Encyclopedia—Volume 6/Test Methods

A Technomic Publishing Company book
Bibliography: p.

Library of Congress Card No. 89-51098
ISBN No. 87762-704-5

C O N T E N T S

FOREWORD

The Delaware Composites Design Encyclopedia provides users with basic knowledge about the design and analysis of composite materials and structures. The six-volume indexed set is an ongoing series to which new volumes will be added as new needs arise and new knowledge is gained about this rapidly growing field of composites. Unlike many other guides on this subject, the encyclopedia emphasizes the underlying fundamental science base—documenting the high-quality research in the various disciplines that contribute to the field—in addition to engineering solutions to specific problems. It is intended for use by engineers, materials scientists, designers, and other technical personnel involved in the applications of composite materials to industrial products.

The material contained in the encyclopedia was written by international experts in the field and compiled at the University of Delaware's Center for Composite Materials (CCM). Established in 1974, CCM began its University-Industry Consortium, "Applications of Composite Materials to Industrial Products," in 1978 to meet the needs of the aerospace, automotive, electronics, and consumer products industries. Now regarded as an international leader in composites research and education, the Center continues to be supported by the Consortium, the United States Army Research Office-University Research Initiatives Program, other federal agencies, and the State of Delaware.

The first version of the *Delaware Composites Design Encyclopedia* was published in 1981 and offered as a special benefit to Consortium members. It has since grown in size and scope and is now being offered to the composites community at large for the first time. The current volumes cover the following subjects:

Volume 1— Mechanical Behavior and Properties of Composite Materials
Volume 2— Micromechanical Materials Modeling
Volume 3— Processing and Fabrication Technology
Volume 4— Failure Analysis of Composite Materials
Volume 5— Design Studies
Volume 6— Test Methods
Index to Volumes 1–6

Dr. John W. Gillespie, Jr. and Dr. Leif A. Carlsson, review editors of the *Delaware Composites Design Encyclopedia*, have incorporated the most recent references relevant to the subject matter to maintain the highest quality, up-to-date encyclopedia of this type.

Volume 6 is authored by international experts R. Byron Pipes, Robert A. Blake, John W. Gillespie, Jr. and Leif A. Carlsson. Pipes reviews test methods (ASTM Standards and Guides) for characterizing constituent properties, composite thermomechanical properties, and physical properties. The influence of environmental effects on properties is also discussed. Delamination represents one of the most prevalent life-limiting failure modes in laminated composite structures. Blake describes ultrasonic nondestructive evaluation techniques and equipment used to assess material quality and detect flaws that affect performance. Numerous examples are documented for various types of composite materials that complement subject matter on the basic principles of the ultrasonic method. Gillespie and Carlsson introduce interlaminar fracture mechanics concepts as background information on test methods for mode I, mode II, and mode III interlaminar fracture characterization of laminated composites. Testing issues and specimen design considerations are emphasized to provide a fundamental understanding of the underlying mechanics principles governing these important material descriptors. Cumulatively, engineers and materials scientists will find this information on test methods essential for materials evaluation and development, quality control assessment of manufacturing techniques, and the creation of material property data bases for the design, analysis, and applications of composite materials to industrial products.

With the completion of the first six volumes of the *Delaware Composites Design Encyclopedia*, I wish to take this opportunity to acknowledge the outstanding contributions of the production team. I wish to personally extend my thanks to Diane Kukich of the University of Delaware Center for Composite Materials for her invaluable assistance in the meticulous editing of galleys and page proofs that enhanced the quality and ensured the timely production of the series. I wish to acknowledge Technomic Publishing Company, which I consider the world's premier composites publisher, for their outstanding contributions to this project. It has been a pleasure working with Michael Margotta, Kier Finlayson, Anthony Deraco, and Kimberly Pugh Martin. I also express my gratitude to Steve Spangler and Eileen Bozarth and all of the other employees working diligently behind the scenes at Technomic that I did not have the pleasure to work with directly.

As new needs arise and new knowledge is gained about the rapidly growing field of composites, the Center for Composite Materials, in collaboration with Technomic Publishing Company, will continue to update and add new volumes written by international experts to this ongoing encyclopedia series.

John W. Gillespie, Jr., Ph.D.
Review Editor and Assistant
 Director for Research
Center for Composite Materials
University of Delaware
Newark, Delaware 19716 USA

DELAWARE COMPOSITES
DESIGN ENCYCLOPEDIA

SECTION 6.1

Test Methods

The characterization of fiber reinforced composites requires the development of a broad range of information that is largely a function of the end use application. However, for an arbitrary application, the following properties should be evaluated for a given material system [1]:

Modulus
- longitudinal modulus (tension and compression)
- transverse modulus (tension and compression)
- longitudinal shear modulus
- major Poisson's ratio (tension and compression)
- minor Poisson's ratio (tension and compression)

Strength
- longitudinal strength (tension and compression)
- transverse strength (tension and compression)
- longitudinal shear strength

Strain
- longitudinal strain to failure (tension and compression)
- transverse strain to failure (tension and compression)
- longitudinal shear strain to failure

Thermal Properties
- coefficient of longitudinal thermal expansion
- coefficient of transverse thermal expansion
- longitudinal thermal conductivity
- transverse thermal conductivity
- specific heat capacity

Moisture Content
- coefficient of longitudinal moisture expansion
- coefficient of transverse moisture expansion
- longitudinal moisture diffusion coefficient
- transverse moisture diffusion coefficient
- maximum moisture content

Impact Energy
- longitudinal impact energy absorption (tension and compression)
- transverse impact energy absorption (tension and compression)
- longitudinal shear impact energy absorption
- Charpy impact (longitudinal and transverse)
- Izod impact (longitudinal and transverse)

Creep
- creep compliance
- relaxation modulus
- longitudinal creep rate (tension and compression)
- transverse creep rate (tension and compression)
- longitudinal shear creep rate

Fatigue
- longitudinal fatigue (tension and/or compression)
- transverse fatigue (tension and/or compression)
- longitudinal shear fatigue

Storage Modulus
- longitudinal storage modulus (tension and compression)
- transverse storage modulus (tension and compression)
- shear storage modulus
- longitudinal loss tangent (tension and compression)
- transverse loss tangent (tension and compression)
- longitudinal shear loss tangent

Electrical Properties
- longitudinal dielectric constant
- transverse dielectric constant
- longitudinal electrical conductivity
- transverse electrical conductivity

Physical Properties
- density
- surface hardness
- wear resistance
- lubricity

Environmental Resistance
- chemical resistance
- resistance to ultraviolet

Many of these properties are strong functions of temperature and moisture content; therefore, each of these properties must be characterized over the temperature and moisture content range of interest.

In the following subsections, the details of test methods that have been developed to measure these properties will be discussed. Emphasis will be placed on those methods that yield intrinsic material properties. A listing of the ASTM test methods for determination of certain of the aforementioned properties is given here:

Mechanical Testing
- D638-77a, Tensile Properties of Plastics
- D695-77, Compressive Properties of Rigid Plastics
- D2344-76, Apparent Horizontal Shear Strength of Reinforced Plastics by Short Beam Method
- D3039-76, Tensile Properties of Oriented Fiber Composites
- D3518-76, In-Plane Shear Stress-Strain Response of Unidirectional Reinforced Plastics
- D790-71 (reapproved 1978), Flexural Properties of Plastics and Electrical Insulating Materials
- D3410-75, Test for Compressive Properties of Oriented Fiber Composites

Fatigue
- D3479-76, Tension-Tension Fatigue of Oriented Fiber Resin Matrix Composites
- D671-71 (reapproved 1978), Flexural Fatigue of Plastics by Constant-Amplitude-of-Force

Impact
- D256-78, Impact Resistance of Plastics and Electrical Insulating Materials
- D1822-68 (reapproved 1973), Tensile-Impact Energy to Break Plastics and Electrical Insulating Materials
- D3029-78, Impact Resistance of Rigid Plastic Sheeting or Parts by Means of Tup (Falling Weight)
- Notched Bar Impact Testing of Metallic Materials

Creep
- D2990-77, Tensile, Compressive, and Flexural Creep and Creep-Rupture of Plastics
- D2991-71 (reapproved 1978), Stress-Relaxation of Plastics

Physical Properties
- D792-66 (reapproved 1975), Specific Gravity and Density of Plastics by Displacement
- D1505-68 (reapproved 1975), Density of Plastics by the Density-Gradient Technique
- D2734-70 (reapproved 1976), Void Content of Reinforced Plastics
- D3355-74, Fiber Content of Unidirectional Fiber/Polymer Composites

Thermal Properties
- D648-72 (reapproved 1978), Deflection Temperature of Plastics under Flexural Load

Thermal Expansion
- D696-70 (reapproved 1978), Coefficient of Linear Thermal Expansion of Plastics
- E228-71, Linear Thermal Expansion of Rigid Solids with a Vitreous Silica Dilatometer

Thermal Conductivity
- C177-76, Steady State Thermal Transmission Properties by Means of the Guarded Hot Plate

Electrical Properties
- D149-75, Dielectric Breakdown Voltage and Dielectric Strength of Electrical Insulating Materials at Commercial Power Frequencies
- D150-78, A-C Loss Characteristics and Permittivity (Dielectric Constant) of Solid Electrical Insulating Materials

Wear Resistance
- D673-70 (reapproved 1976), Mar Resistance of Plastics
- D1242-56 (reapproved 1975), Resistance of Plastic Materials to Abrasion

Chemical Resistance
- C581-74 (reapproved 1978), Chemical Resistance of Thermosetting Resins Used in Glass Fiber Reinforced Structures
- D543-67 (reapproved 1978), Resistance of Plastics to Chemical Reagents

For consistency with the terminology associated with more conventional structural materials, the following existing definitions of statistical allowables are also recommended for use with advanced composite constituents and laminates:

- *A*-Basis—The *A* mechanical property value is the value above which at least 99 percent of the population of values is expected to fall, with a confidence of 95 percent.

- *B*-Basis—The *B* mechanical property value is the value above which at least 90 percent of the population of values is expected to fall, with a confidence of 95 percent.
- *S*-Basis—The *S* mechanical property value is the minimum value specified by the governing specification.
- Typical Basis—The typical property value is an average value. No statistical assurance is associated with this value.

Usually, only tensile, ultimate, and yield strengths in a specified testing direction are determined in such a manner that they can be termed *A* or *B* values in accordance with the foregoing definitions. Likewise, usually only tensile, ultimate, and yield strengths and elongation are specified in the governing specifications; they can be termed *S* values. However, scaling procedures have been established by which other property values can be computed with approximately the same assurance levels as the *A*, *B*, or *S* values for tensile, ultimate, and yield strengths.

Where the quantity of data permits meaningful statistical evaluation, *A*-basis and *B*-basis allowable strength values should be computed. Other properties, such as elastic modulus, Poisson's ratio, and physical properties, are normally presented as typical (average) values. In general, fatigue and creep strength are typical values unless otherwise indicated.

6.1.1 Constituent Test Methods

It is often desirable to obtain constituent properties as part of the total characterization of composite materials. Tensile properties of fiber and matrix are of particular interest. Determination of matrix properties may be hindered for certain resin systems by the inability to fabricate castings. This problem can usually be overcome, however, by utilizing thin film specimens. Many of the physical property characterization methods for constituents are essentially the same as for composites. As a result, these methods will be covered under subsection 6.1.2.

Single-Filament Tensile Properties

This test method is used to determine strength, modulus, and failure strain of single filaments. The method consists of applying constant strain rate loading to a single-filament centerline mounted on special slotted tabs (see Figure 6.1-1). This method is discussed in detail under ASTM D3379-75 and is limited to fibers with a modulus above 21 GPa (3 Msi). The fixed gage length of the fibers, *L*, must be at least 2,000 times the nominal filament diameter. The gripping system used in conjunction with the tabs shown in Figure 6.1-1 must be designed such that axial alignment may be easily accomplished without damaging the filament. After the specimen is mounted in the test machine, the center section of the tab is burned or cut away to allow for filament elongation.

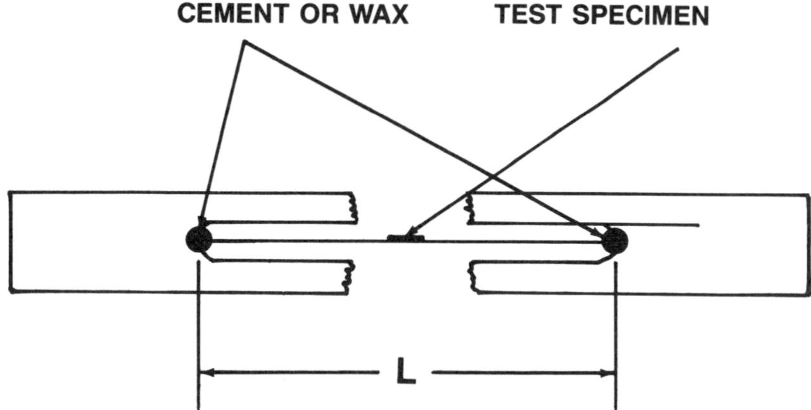

FIGURE 6.1-1. Single-filament test method.

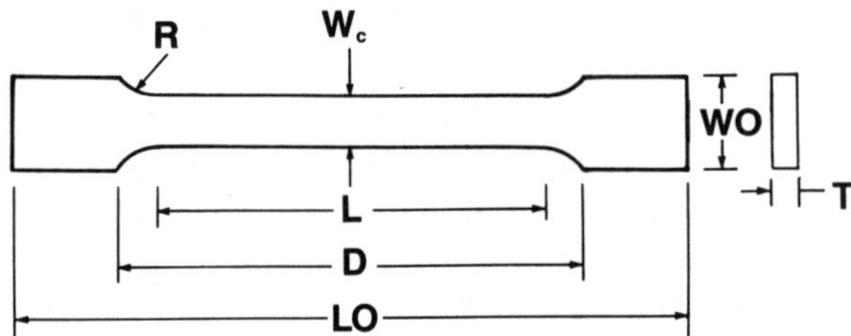

W_c = WIDTH OF NARROW SECTION
L = LENGTH OF NARROW SECTION
WO = OVERALL WIDTH, MINIMUM
LO = OVERALL LENGTH, MINIMUM
D = DISTANCE BETWEEN GRIPS
R = RADIUS OF FILLET
T = THICKNESS

FIGURE 6.1-2. Schematic of neat resin tensile test specimen.

Filament cross-sectional area, A, is determined by planimeter measurements of a representative number of filament cross sections as displayed on highly magnified photomicrographs. For fibers with large cross-sectional variations, such as graphite, a large number of sample measurements is necessary in order to obtain an accurate average value of the cross-sectional area. Other methods of cross-sectional area determination include optical gages and image-splitting microscopes.

The test procedure consists of loading the specimen to failure at a constant cross-head rate and recording a load-displacement curve. The filament strength, σ_f, is simply the maximum load, P_{max}, divided by the average cross-sectional area:

$$\sigma_f = \frac{P_{max}}{A} \qquad (6.1\text{-}1)$$

An apparent fiber modulus, E_a, can be determined by

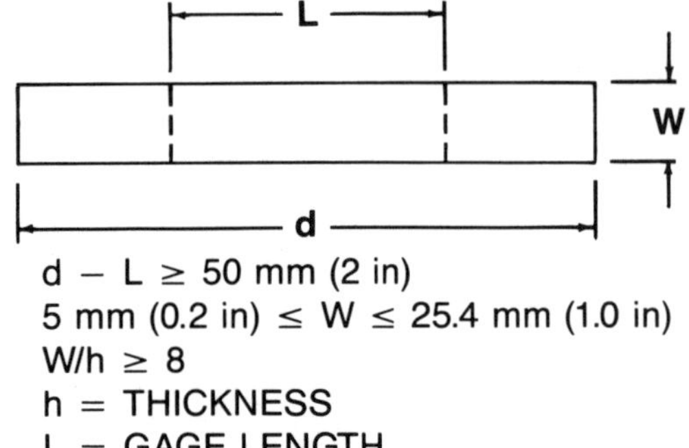

d − L ≥ 50 mm (2 in)
5 mm (0.2 in) ≤ W ≤ 25.4 mm (1.0 in)
W/h ≥ 8
h = THICKNESS
L = GAGE LENGTH

FIGURE 6.1-3. Schematic of alternate neat resin tensile test specimen.

measuring the slope of the initial straight line portion of the load-displacement curve:

$$E_a = \frac{PL}{Au} = \frac{L}{AC_a} \qquad (6.1\text{-}2)$$

where P and u are the load and cross-head displacement, respectively, associated with the slope of the initial straight line portion of the load-displacement curve on the recording chart, which quantifies the apparent compliance C_a. The term *apparent modulus* is used because cross-head travel, due to system compliance, is not a true measure of filament strain. This difficulty can be overcome by substituting the system compliance, C_s, from the measured compliance. The fiber stiffness is defined as

$$E_f = \frac{PL}{A(u - u_s)} = \frac{L}{A(C_a - C_s)} \qquad (6.1\text{-}3)$$

The system compliance can be determined by measuring the apparent compliance for different gage length specimens. In particular, a best fit straight line to a plot of measured values of C_a as a function of L will yield C_s as the y-intercept. The total deformation of the system, u_s, is defined as

$$u_s = C_s P \qquad (6.1\text{-}4)$$

The fiber strain at failure, ϵ_f, is simply

$$\epsilon_f = \frac{u_{max} - C_s P_f}{L} \qquad (6.1\text{-}5)$$

where u_{max} is the maximum cross-head displacement at the failure load, P_f, as determined from the recording chart. A more accurate strain measurement can be obtained by optical techniques.

Polymeric Matrix Tensile Properties

For resin systems that can easily be processed in thick sheets, a standard dogbone type specimen, as shown in Figure 6.1-2, is used for tensile property determination. The exact dimensions of the dogbone specimen depend on thickness. Details are spelled out in ASTM D638-72. The specimens may be prepared by machining or die cutting them from materials in sheet or plate form, or they may be prepared by molding.

Strain measurements can be obtained by using an extensometer. If a measure of Poisson's ratio is desirable, longitudinal and transversely oriented strain gages can be employed, provided the specimen is sufficiently thick.

Resin systems that cannot be readily fabricated in thick sheets can often be fabricated into thin sheets or films. Tensile tests can be performed on strips of such material cut from sheets or obtained from molds. The details of this test method are discussed in ASTM D882-73. The method is applicable to specimens with a thickness, h, less than 1.0 mm (0.04 in). The specimen is shown in the schematic of Figure 6.1-3, where the test gage section is denoted by L. For specimens that are not too thin, an extensometer can be used to measure strain. For flexible specimens, optical techniques may be used for strain measurement.

It should be noted that a dogbone type film specimen can be utilized by molding it to the desired dimensions.

6.1.2 Physical Property Test Methods

In this subsection, density measurement, fiber volume content, and measurement of expansional strains are discussed. The method for density measurement is applicable to both fibers and resins as well as composites, whereas methods for determining thermal and moisture expansion coefficients are limited to polymeric matrix materials and composites.

Density

Density for polymeric matrix resins and fiber reinforced composites may be determined by measuring the difference between the weight of a specimen in air and in water. The volume of the specimen must be at least 1 cm³ (0.06 in³), and the surface edges should be smooth.

The test procedure consists of measuring the specimen weight in air, then weighing the specimen while it is suspended on a wire and immersed in a container of water. For specimens with a density less than 1 g/cm³, a sinker must be attached to the wire. It is also necessary to measure the weight of the completely immersed sinker and partially immersed wire. The density, ϱ, is then determined from the relationship

$$\varrho = \frac{(0.9975)a}{(a + w - b)} \qquad (6.1\text{-}6)$$

where a is the weight of the specimen in air, b is the weight of specimen and sinker completely immersed and of the wire partially immersed, and w is the weight of the totally immersed sinker and partially immersed wire. This test method is covered by ASTM D792-66.

A similar procedure may be used for measuring the density of single filaments. The liquid chosen, however, must be capable of completely wetting the fiber.

Fiber Volume Fraction

Two methods are available for determining fiber volume content. One method involves digestion of the matrix by a liquid medium with the weight of the remaining fiber being used in conjunction with the fiber and composite densities to determine volume percent of fiber. A second method involves a simple determination of the composite density. If the composite is essentially free of voids, then the composite density can be used in conjunction with the fiber and matrix densities to calculate fiber volume content [2].

The digestion method is applicable to both polymeric and metal matrix composites. Choice of liquid for digestion is governed by the specific matrix and fiber. For example, epoxy resins are readily digested by sulfuric acid, but graphite fibers are attacked by sulfuric acid. As a result, hot nitric acid is used in conjunction with graphite/epoxy composites. Graphite is also attacked by nitric acid if it is allowed to remain in the acid for too long a time period. The attack, however, is much less severe than in the case of sulfuric acid. Thus, the liquid of choice for digestion must be compatible with the fiber while it does an efficient job of digesting the matrix. More details concerning the choice of liquid for digestion can be found in ASTM D3171-76 for polymeric matrix composites and in ASTM D3553-76 for metal matrix composites.

$$V_f = \frac{(W_f/\varrho_f)}{(W_c/\varrho_c)} \qquad (6.1\text{-}7)$$

where W_f, W_c, ϱ_f, and ϱ_c are the fiber weight, composite weight, fiber density, and composite density, respectively. If the matrix density, ϱ_m, is known, the void volume fraction, V_v, can be calculated from the relationship

$$V_v = 1 - \left[\frac{W_f/\varrho_f - (W_c - W_f)/\varrho_m}{W_c/\varrho_c} \right] \qquad (6.1\text{-}8)$$

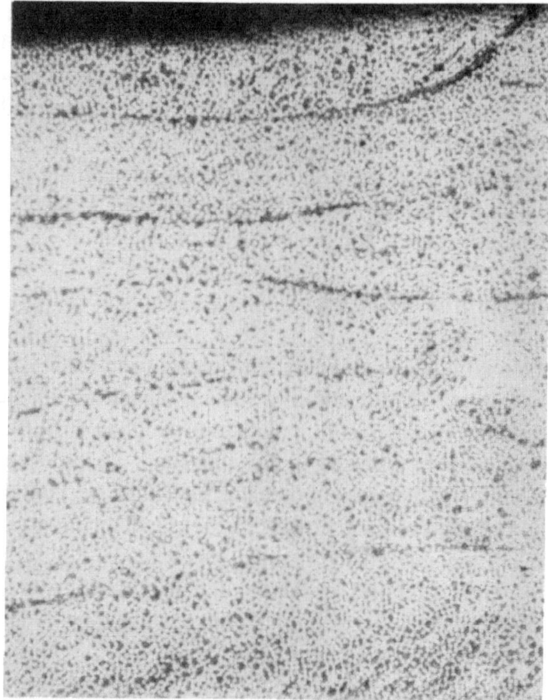

(a) VOID FREE

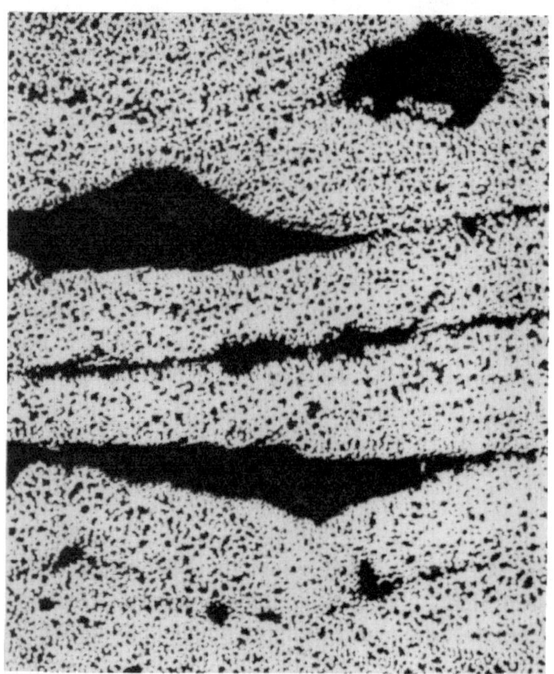

(b) WITH VOIDS

FIGURE 6.1-4. Micrograph of graphite/epoxy.

If some weight loss by the fiber is suspected, a correction can be made by exposing a bundle of fibers (of approximately the same weight as the fibers in the composite specimen) to the acid for the same length of time as the composite exposure. Fiber weight loss is then determined by measuring the difference between the fiber weight before and after exposure. Only weight losses greater than 0.5% are considered significant. Caution should be exercised on this correction procedure, however, because the fibers in the composite are not exposed to direct contact with the digesting medium during the full time the resin is being dissolved.

Metal matrix composites are essentially void-free because of the extreme pressure used in processing them. In addition, many polymeric matrix composites are fabricated with void contents less than 1%. In such cases, it is assumed that the void content is zero, and the fiber volume fraction can be determined from the densities of the fiber, composite, and matrix through the relationship

$$V_f = \left(\frac{\varrho_c - \varrho_m}{\varrho_f - \varrho_m} \right) \tag{6.1-9}$$

Void content can be checked from photomicrographs of composite cross sections. Typical cross sections are shown in Figure 6.1-4 for a graphite/epoxy unidirectional composite. The cross section in Figure 6.1-4(b) shows a large void content (dark spots are voids), while the cross section in Figure 6.1-4(a) is virtually void-free.

It should be noted that many fibers, such as graphite, tend to have a large variation in density. In such cases, both of the procedures discussed for determining fiber volume content yield only approximate results.

Coefficient of Thermal Expansion

Measurement of the linear coefficient of thermal expansion simply involves measuring the dimensional change of a specimen over a temperature range. To verify linearity, it is necessary to plot thermal strain as a function of temperature [1,2]. Typical results are shown in Figure 6.1-5 for the transverse thermal expansion coefficient, α_2, of an HTS unidirectional graphite/epoxy composite [3]. Transverse strain, ϵ_2, was measured using a strain gage. The linear coefficient of thermal expansion is the slope of the ϵ_2 versus T plot over the initial straight line portion of the curve. A factor is added to compensate for the difference between the thermal expansion coefficient of the gage and the composite [1,2]. For the example in Figure 6.1-5, the curve is linear up to a temperature that is close to the composite cure temperature, where the material response displays considerable viscoelasticity.

Other appropriate devices, such as an extensometer or a laser dilatometer [4], may be used to measure thermal strain. The extensometer technique can also be used on polymeric resins that can be fabricated into thick sheets. A detailed procedure for cast resins can be found in ASTM D696-70. Since absorbed moisture induces swelling in many resin systems, the specimens used for determining coefficient of thermal expansion should be initially dry.

Coefficient of Moisture Expansion

Measurement of the coefficient of moisture expansion involves measuring the dimensional change of a specimen over a range of moisture weight gains. The experiment may be performed by placing a specimen in a water bath and measuring swelling strain as a function of weight gain. Difficulty arises in the determination of strain because conventional strain gage adhesives are attacked by moisture; therefore, the accuracy of strain gage measurements is reduced with exposure time.

Hahn and Kim [3] used a micrometer and a caliper to measure swelling strains on graphite/epoxy composites. The gage length of the micrometer was 2.54 cm with a resolution of 0.00254 mm, and the gage length of the caliper was 20.3 cm with a resolution of 0.0127 mm.

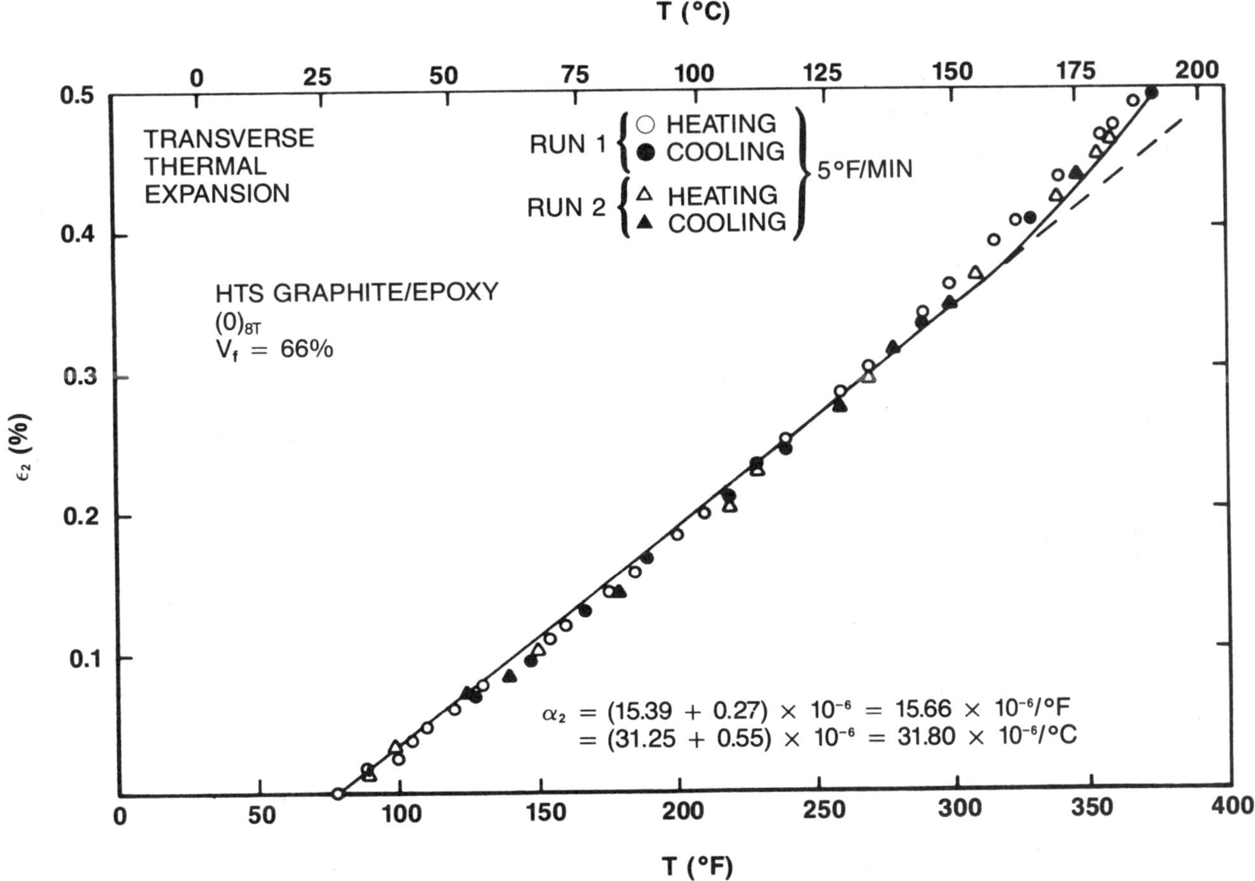

FIGURE 6.1-5. Transverse thermal expansion.

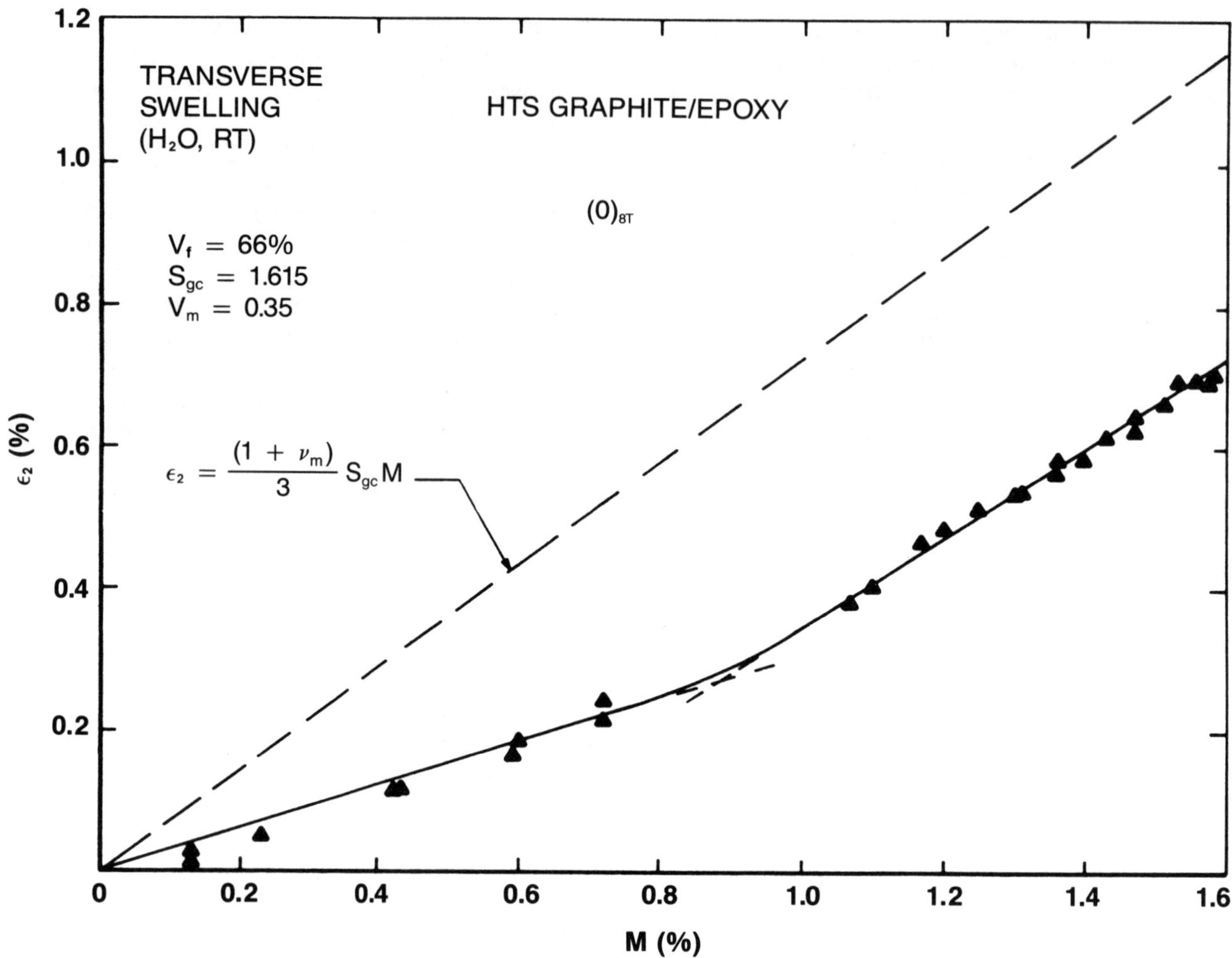

FIGURE 6.1-6. Transverse moisture swelling results.

Strain measurements were taken from an average of three readings for in-plane strains and an average of four readings for through-the-thickness strains. This technique can be used on both neat resins and composites.

In the case of unidirectional composites, most of the swelling occurs transverse to the fibers, whereas, for laminated composites, most of the swelling occurs through the thickness.

It has been noted by Hahn [5] that moisture absorption in graphite/epoxy composites is associated with a swelling threshold described by the relationship

$$\epsilon_i^H = 0, \; c \leq c_0$$

$$\epsilon_i^H = \beta_i(c - c_0), \; c_0 < c \qquad (6.1\text{-}10)$$

where c_0 is a threshold moisture concentration. Such a threshold moisture level may be associated with water filling microvoids or cracks.

Because of the relationship given by Equation (6.1-10), care must be exercised in measuring β_i from a plot of swelling strain versus total weight gain.

Consider the case of a unidirectional graphite/epoxy composite in which the in-plane dimensions are large compared to the thickness such that the moisture diffusion is essentially one dimensional in nature. A plot of the average transverse strain, ϵ_2, versus total percent weight gain, M, yields the result shown in Figure 6.1-6. The initial portion of this curve is associated with a moisture gradient that has not exceeded c_0 throughout the laminate. At some time, $t = t_0$, $c \geq c_0$ throughout the composite, and the swelling strains will be linear with the total weight gain. This relation can be illustrated using Equation (6.1-10) as follows:

$$\epsilon_2(t) = \frac{2}{h} \int_0^{h/2} \epsilon_2^H dz = \beta_2 \varrho_c [M(t) - M_0], \quad t_0 \le t$$

$$(6.1\text{-}11)$$

where ϵ_2 denotes the average strain through the thickness as measured in the experiment. Thus, if S_2 is the slope of the secondary straight line portion of Figure 6.1-6,

$$\beta_2 = \frac{S_2}{\varrho_c} \qquad (6.1\text{-}12)$$

The value of M_0 can be obtained by extending the secondary straight line portion of the curve through the M-axis.

The dotted line in Figure 6.1-6 assumes that all of the absorbed moisture is translated into a matrix volume change. Such behavior does occur after the threshold moisture gain has been attained throughout the composite. Using micromechanics [5],

$$\epsilon_2(t) = \frac{(1 + \nu_m)}{3} S_{gc}[M(t) - M_0], \quad t_0 \le t \qquad (6.1\text{-}13)$$

where S_{gc} is the specific gravity of the composite. Combining Equations (6.1-11) and (6.1-13) yields

$$\beta_2 = \frac{(1 + \nu_m)}{3\varrho_w} \qquad (6.1\text{-}14)$$

where ϱ_w is the density of water.

For some composites and neat resins, c_0 may vanish, and the swelling strains will be linear with M up to full saturation.

6.1.3 Environmental Testing

The glass transition temperature, T_g, of a polymer is defined as the temperature below which it is hard and above which it is soft. The hard polymer is a glasslike material, while the soft polymer varies from a rubbery material to a viscous liquid [6]. For cross-linked amorphous polymers such as epoxy resins, T_g is governed by

the cure or postcure temperature. Thus, T_g can be increased for these systems by raising the cure or postcure temperature.

Viscoelastic Properties

From a practical standpoint it is more appropriate to discuss a glass transition temperature region rather than a single glass transition temperature because the change from a hard polymeric material to a soft material takes place over a temperature range. This range, as illustrated in Figure 6.1-7, is characterized by a decreasing modulus E_R with increasing temperatures above T_g. Thus, viscoelastic properties are a concern over a range of temperatures below as well as above T_g [7].

Effect of Moisture

It is well recognized [6] that the T_g of a polymer can be lowered by mixing with it a miscible liquid (diluent) that has a lower glass transition temperature than the polymer. This process is referred to as *plasticization*. Many polymers, such as epoxy resins, absorb moisture from high humidity environments [8–10]. Thus, moisture is a plasticizer for such systems, producing a lower value of T_g. Similar effects are observed in composites [7,9].

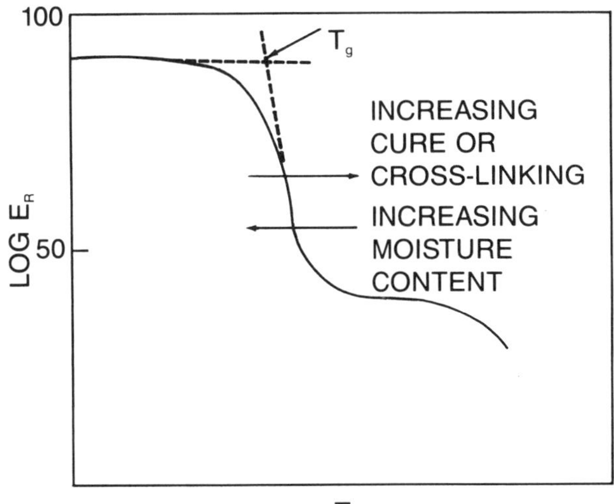

FIGURE 6.1-7. Glass transition temperature.

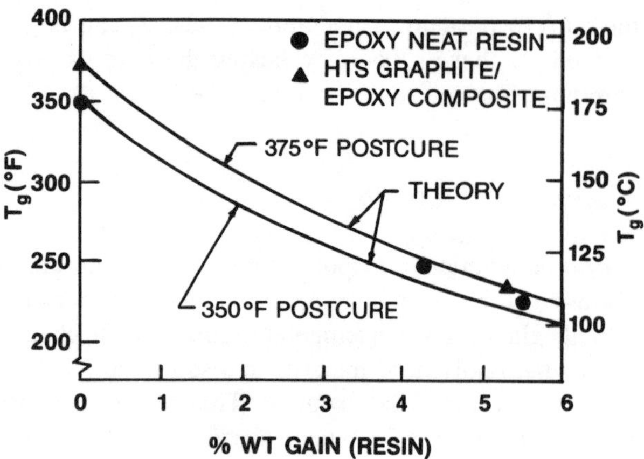

FIGURE 6.1-8. Glass transition temperature as a function of absorbed moisture in resin.

The lowering of T_g with increasing moisture content is illustrated in Figure 6.1-8 for both an epoxy resin and a derived graphite/epoxy composite. Theoretical results are based on the Bueche-Kelley theory for T_g of a polymer-diluent system [11]. It should be noted that the dry T_g is higher for the composite than for the resin. This is due to the fact that the resin was not postcured, yet the composite was postcured at 191°C (375°F).

Measurement of T_g

A number of methods are available for measuring the T_g of polymers [12]. A procedure that appears to give consistent results and is easily adaptable for both resins and composites is the heat distortion temperature (HDT) method. This test consists of the three-point loading of a flexure specimen surrounded by an increasing temperature environment. A fixed load is applied at the center of the beam and the deflection measured as a function of temperature. The glass transition region is also characterized by a rapid increase in the coefficient of thermal expansion [12]. The choice of actual T_g is somewhat arbitrary; however, the point where the slope of the deflection temperature curve becomes positive yields very consistent results. The test procedure for polymeric materials is described in ASTM Standard D648-72.

For composite materials the test procedure is essentially the same as for polymers. Typical results, test geometry, and applied load are shown in Figure 6.1-9 for a matrix dominated specimen having [90/±45]$_s$ lay-up. The specimen is placed on edge so that the bending deformation takes place in-plane rather than through the thickness. An oil bath surrounds the specimen as a heat transfer medium and also slows down the drying out process for wet T_g measurements.

Because of the lowering of the resin glass transition temperature, there is considerable interest in determining the mechanical properties of composite laminates at various temperatures in the presence of moisture [10]. Such experimental evaluation requires a knowledge of the moisture diffusion process in order to perform moisture conditioning and to determine moisture content and distribution during elevated temperature tests.

Moisture Measurements

Diffusion experiments consist of measuring weight gain as a function of time for a constant temperature and humidity exposure. The concentration of moisture on the surface of an exposed specimen is a function of relative humidity; as a result, the equilibrium moisture concentration is also a function of humidity [10]. In most engineering applications, moisture diffusion is through a large surface area with very few edges. As a result, the diffusivity through the thickness, $\bar{d}_{33}$, is of primary interest. For such a case, thin composite specimens are utilized in order to approximate a one-dimensional diffusion process. If the use of a thin composite is not practical, the edges can be sealed with foil or an appropriate coating to retard diffusion.

Diffusion coefficients that are a function of temperature appear to follow an Arrhenius-type relationship of the form [10,13,14]

$$d = d_0 \exp\left(-E_d/RT\right) \qquad (6.1\text{-}15)$$

where d_0 is a constant, E_d is the activation energy for diffusion, R is the universal gas constant, and T is temperature as measured on the Kelvin scale. Thus, the diffusion process must be characterized for different temperatures as well as relative humidities.

All specimens for absorption should be preconditioned in a vacuum oven until a near equilibrium weight is obtained. This procedure will assure initially dry specimens. The temperature of the vacuum oven should not exceed 200°F (93°C). The drying temperature should never exceed T_g. Thus, for resins with low values of T_g, the vacuum oven temperature may have to be less than 200°F (93°C).

After preconditioning, dry specimens are placed in an environmental chamber under constant temperature and constant humidity. If a chamber in which relative humidity can be controlled is not available, specimens can be placed in a container of water inside an oven, and the underwater condition will approximate 100% relative humidity. If other humidity conditions are desired, a humid desiccator can be employed in conjunction with the oven [14]. Specimens should be allowed to cool for a short period of time before being weighed. For specimens placed underwater, any surface moisture should be carefully wiped off before weighing. The ideal exposure, obviously, would consist of a chamber in which weight gain could be monitored continuously without specimen removal. The weight gain process is continued until an apparent equilibrium is reached.

The data reduction procedure consists of plotting normalized moisture content, M/M_e, as a function of $\sqrt{t}$ for each exposure temperature [15]. Since the data are normalized by the equilibrium concentration, different humidity conditions can be plotted on the same curve. For specimens of small thickness, the data should approximate a straight line in the region $0 \le M/M_e \le 0.6$. For thicker laminates, the diffusion equations will have to be evaluated to assess the upper limit of M/M_e that produces a straight line.

The slope of the straight line portion of the M/M_e versus $\sqrt{t}$ plot is used in conjunction with the following equations to determine the through-the-thickness diffusion coefficient, d_{33}:

$$\bar{d} = \frac{\pi^2 h}{16}\left(\frac{G}{\sqrt{t}}\right)^2 \qquad (6.1\text{-}16)$$

$$d_{33} = \bar{d}/(1 + \sqrt{R_1} + \sqrt{R_2}) \qquad (6.1\text{-}17)$$

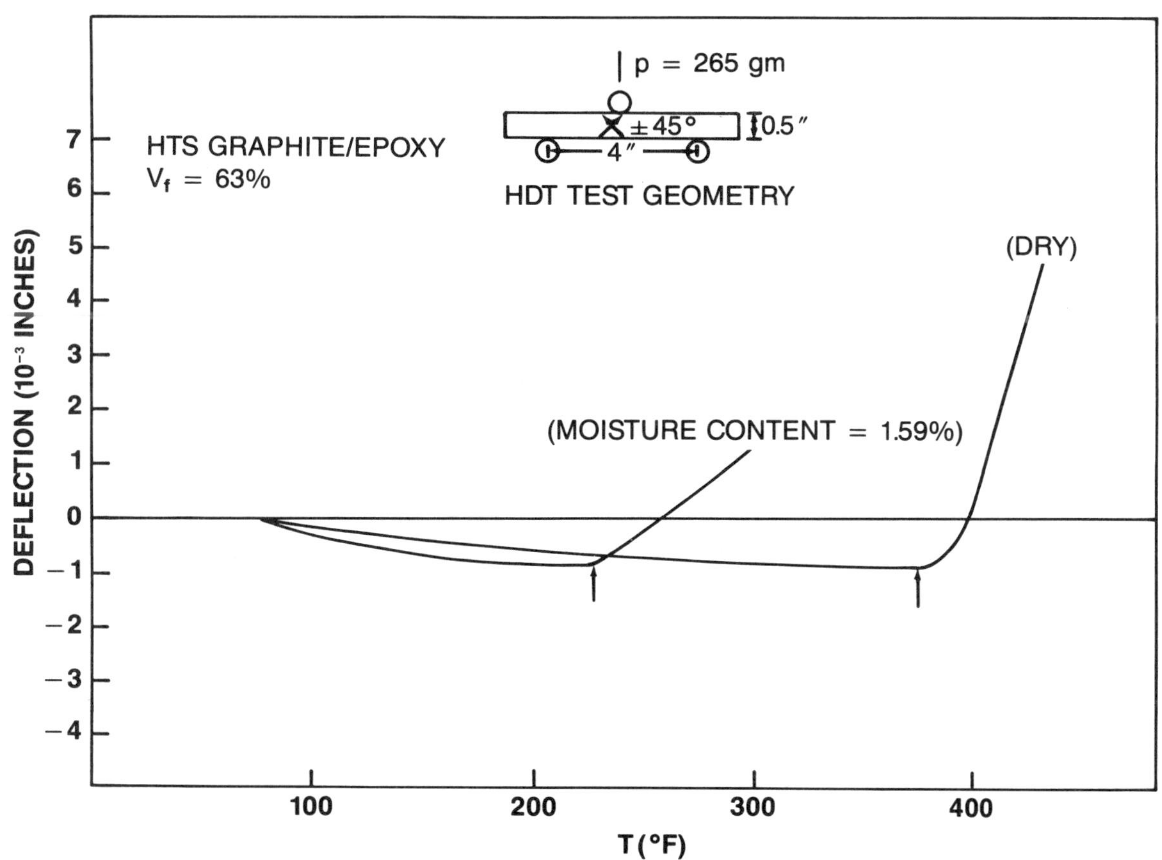

FIGURE 6.1-9. Typical HDT test results for graphite/epoxy.

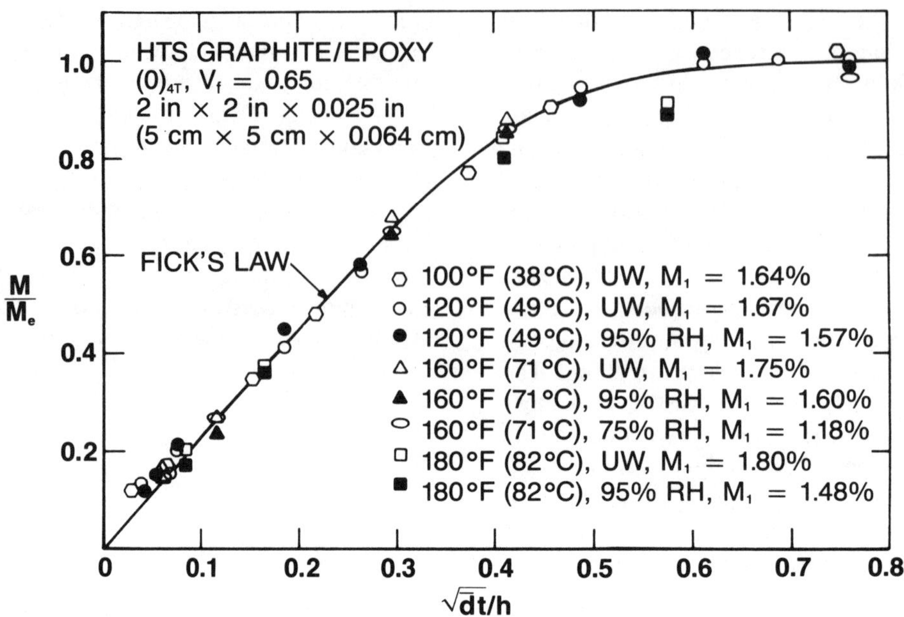

FIGURE 6.1-10. Master plot for graphite/epoxy unidirectional composite [15].

where

$$G = 4\frac{\bar{t}^*}{\pi}$$

$$\bar{t}^* = \frac{\bar{d}t}{h^2}$$

$$R_1 = \frac{h^2\bar{d}_{11}}{a^2\bar{d}_{33}} , \quad R_2 = \frac{h^2\bar{d}_{22}}{b^2\bar{d}_{33}} \qquad (6.1\text{-}18)$$

$$\bar{d}_{ij} = \frac{1}{h}\sum_{k=1}^{n} d_{ij}^k \quad (n \text{ laminae})$$

$$d_{ij}^k = \text{lamina diffusivity components}$$

A plot of M/M_e versus $\sqrt{t^*}$ allows data from different relative humidities and temperatures to be plotted on one master plot. Such a plot is shown in Figure 6.1-10 for a unidirectional graphite/epoxy composite [15]. Alternate principles for data reduction are outlined in reference [10].

If desorption experiments are to be utilized for measuring diffusivity, the procedures as described here require that moisture adsorption must occur until equilibrium is reached prior to running the desorption test. This procedure assures a uniform initial moisture profile. If the initial moisture distribution is not uniform, the data reduction procedures previously described are not valid. For desorption, the equilibrium moisture concentration is zero, and

$$G = \frac{M_0 - M}{M} \qquad (6.1\text{-}19)$$

where M_0 is the initial moisture concentration. Thus, the previously described procedures can be used for desorption except that G, rather than M/M_e, is plotted as a function of $\sqrt{t}$. A more complete discussion of combinations of absorption and desorption experiments can be found in reference [15].

Note that all of the diffusion measurements discussed consider only total weight gain. To date, no acceptable method has been developed for measuring the moisture distribution within the laminate for composite materials.

Mechanical Properties

In most engineering usage of fiber reinforced composites, laminate stacking geometry is chosen such that stiffness and strength are controlled by fiber modulus and strength, respectively. Thus, some matrix softening can be accommodated in such applications without serious consequences. If considerable matrix softening occurs, however, the ability of the resin to support the fiber, along with the ability to transfer load through the matrix to the fibers, is severely reduced. The result is a change in failure mode from filament dominated to matrix dominated. The classical example is that of unidirectional compression, where a significant loss in matrix stiffness leads to local instabilities and a reduction in compressive strength. Thus, any loss in resin T_g due to moisture absorption can lead to a reduction in the useful temperature range of the composite laminate.

Another classic example of a failure mode change is the unidirectional flexure test. This test is commonly used for quality control; zero degree flex strength is considered to be a filament dominated property. For state-of-the-art, high-performance epoxy resins, zero degree dry flex strength is relatively insensitive to temperatures below 149°C (300°F), but it falls off at higher temperatures. Such strength degradation is accompanied by a change in failure mode from filament dominated to matrix dominated [16]. This change of mode is illustrated in Figure 6.1-11 for 4-point loading of a graphite/epoxy unidirectional composite [16].

Thus, it is easily seen from these examples that the characterization of mechanical properties in the pres-

ence of moisture and temperature is of practical concern.

Accelerated Conditioning

In order to minimize preconditioning time, accelerated moisture conditioning techniques are often employed in conjunction with mechanical test specimens [17]. This preconditioning time reduction can be accomplished by increasing exposure temperature (Figure 6.1-12a) and humidity.

If the preconditioning moisture content is less than the value of M_e associated with 100 percent humidity exposure, then exposure at high humidity levels will reduce the conditioning time. For example, if $M_e = 2.0$ percent for 100 percent relative humidity and 1.0 percent for 50 percent relative humidity, then the equilibrium moisture content associated with the lower relative humidity can be obtained by exposure at a humidity level above 50 percent for a shorter period of time (see Figure 6.1-12b). Actual weight gain, M, is shown for examples of both 100 percent and 50 percent relative humidity exposure. It is obvious that considerable time can be saved in obtaining the 1.0 percent weight gain by exposure at 100 percent relative humidity. It should be noted, however, that this procedure yields a very non-

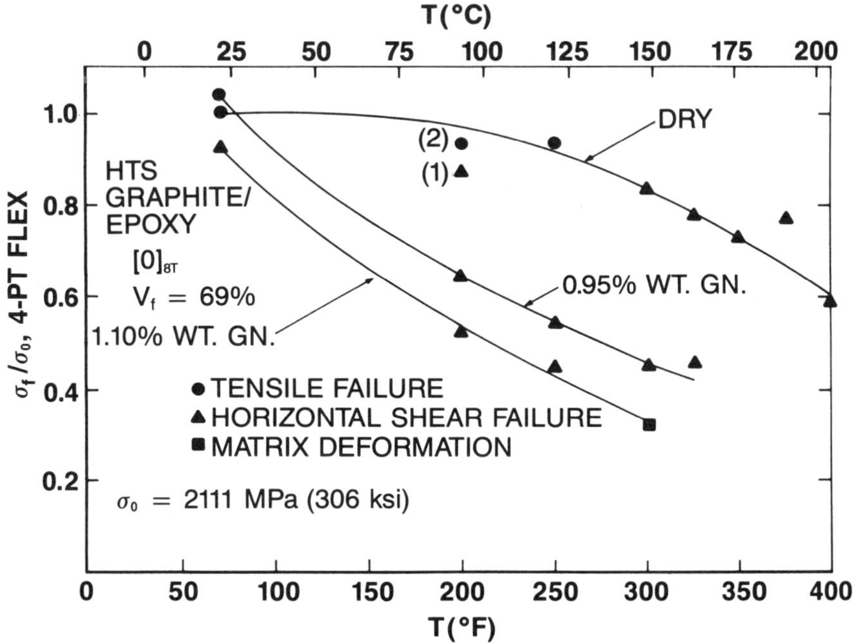

FIGURE 6.1-11. Flexure strength, 4-point loading of graphite/epoxy.

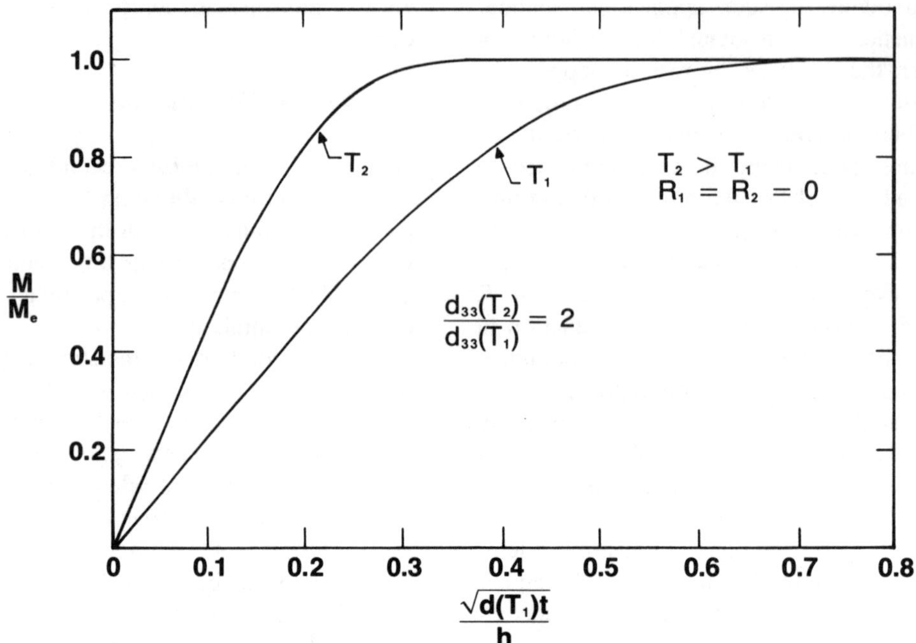

FIGURE 6.1-12a. One-dimensional moisture absorption for temperatures T_1 and T_2.

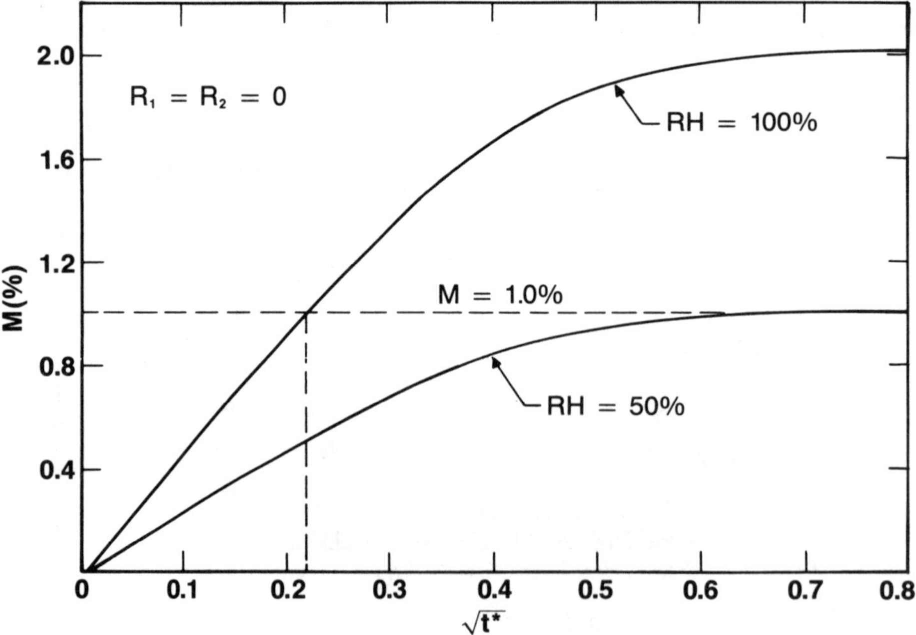

FIGURE 6.1-12b. Time to reach 1.0 percent moisture concentration for two relative humidity exposures.

uniform moisture distribution across the laminate thickness for the 1.0 percent weight gain associated with 100 percent relative humidity. The moisture profiles associated with 1.0 percent weight gain for the two humidity exposures are shown in Figure 6.1-13. Note that, while the total weight gain M is at the desired 1.0 percent level for 100 percent exposure, the outer surface, $z/h = 0.5$, is at 2.0 percent, while the center, $z/h = 0$, is almost dry. Such a gradient causes severe swelling stresses and may yield very misleading results in terms of mechanical properties.

Care should also be taken in high temperature moisture exposure; in particular, temperatures that are close to the wet glass transition temperature of the resin may induce permanent damage to the test specimen in the formation of microcracks. In addition, if large amounts of water associated with high humidity exposure are rapidly forced into the composite by high temperatures, rapid swelling will occur with an increased possibility of inducing cracks. Further discussion on these points can be found in references [14,15].

End Tabs and Strain Gages

Most of the standard adhesives used for strain gages and specimen end tabs require heat and pressure for varying lengths of time. Such procedures induce drying out of preconditioned specimens. Moisture sensitivity of most adhesives precludes bonding prior to moisture conditioning. Silicone rubber and foil have been placed over strain gages in order to isolate them from moisture contact during environmental conditioning. Such a procedure is only partially effective, however, because moisture diffusion from inside the composite to the outer surfaces will eventually penetrate the bondline.

For room temperature specimens, these problems can be overcome by utilizing a room temperature curing, anaerobic adhesive (for example, Eastman 910 or M-Bond 200). Such adhesives can be used for end tabbing and strain gaging after moisture conditioning. For elevated temperature specimens, a high temperature anaerobic adhesive (for example, Loctite 306NF Primer) can be used. This type of adhesive will cure at room temperature and complete its cure during heat-up of the specimen in preparation for testing. This procedure, as applied to strain gages, is fully described in reference [18].

Moisture Control

Very often it is necessary to perform mechanical tests at elevated temperatures in a dry oven that induces drying out of the specimen. Although environmental chambers are available that control humidity below the boil-

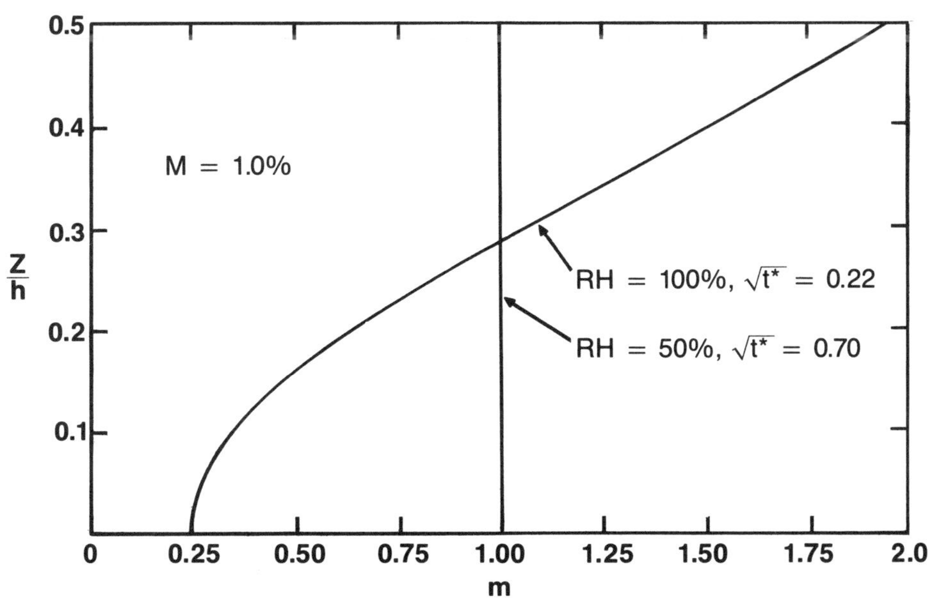

FIGURE 6.1-13. Moisture profiles at 1.0 percent weight gain for two relative humidity exposures.

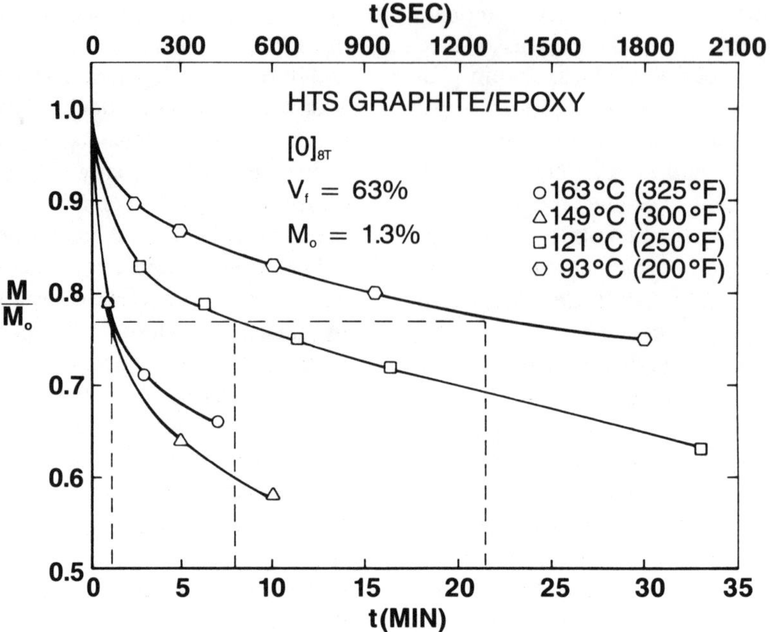

FIGURE 6.1-14. Weight loss as a function of time, graphite/epoxy.

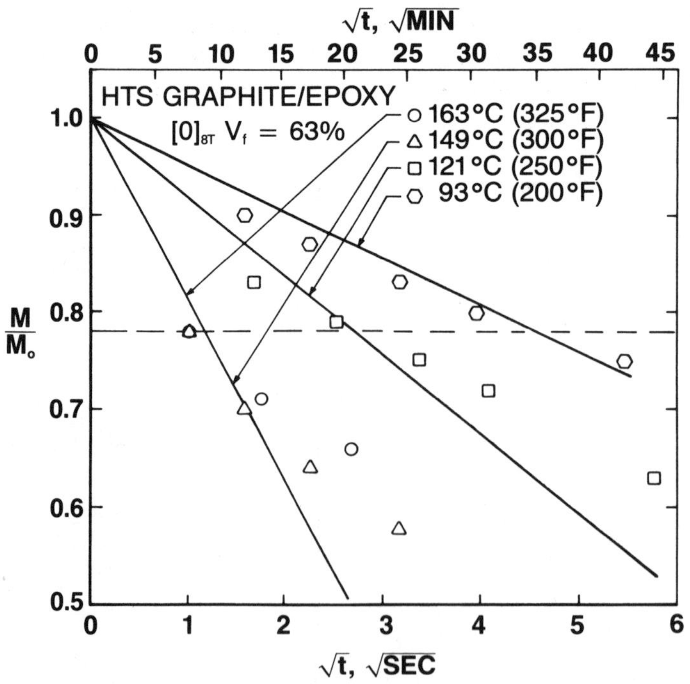

FIGURE 6.1-15. Weight loss as a function of square root time.

ing point of water, higher temperatures with humidity control cannot be obtained without a pressure chamber. Such equipment is not commonly found in the laboratory.

As previously discussed, diffusivity is a function of temperature. Thus, the amount of weight loss associated with elevated temperature tests in a given time period will be a function of temperature. In order to compare the degradation in mechanical properties at various elevated temperatures for a given moisture content, the hold time prior to testing must be adjusted for different temperatures.

Such a procedure was implemented in reference [16] in conjunction with unidirectional flexure specimens. As discussed previously, both the total weight gain and the detailed moisture profile are determined by t^* for constant values or R_1 and R_2. Thus, control of t^* provides a means of controlling the detailed moisture distribution during a drying process. In particular, a plot of M versus $\sqrt{t}$ should yield a straight line for a process in which the diffusion coefficient is constant. From such a plot, a value of t can be chosen for different temperatures such that t^* is a constant, which will assure the same total weight loss and moisture profile for each temperature associated with a given value of M_0.

In order to control the moisture content during the flexure experiments of reference [15], weight loss at various temperatures as a function of time was determined for specimens fabricated from two different material systems. Results are shown in Figure 6.1-14 for HTS graphite/epoxy specimens. The test procedures consisted of preheating the test chamber (including the flexure fixture) to the desired temperature, then placing the specimen in the loading fixture and holding it for a predetermined length of time before loading the specimen. A minimum time of about 90 seconds was required to bring the specimen to temperature equilibrium. The weight loss at all elevated temperatures was set equal to the 163°C (325°F), 90 second weight loss. 163°C was the maximum test temperature. For the HTS graphite/epoxy specimens in Figure 6.1-14, this temperature equilibrium required a seven-minute temperature soak at 121°C (250°F) and a twenty-two-minute soak at 93°C (200°F).

The data in Figure 6.1-14 are plotted in Figure 6.1-15 as a function of $\sqrt{t}$. At 93°C (200°F), an approximate straight line is obtained. For the other elevated temperatures, transient heating effects in conjunction with the shorter times yield results that depart drastically from

linearity, indicating that the moisture profile was different for all elevated temperatures.

Because of these difficulties in controlling weight loss during elevated temperature tests, it is strongly recommended that mechanical testing be carried out in a humidity controlled environment whenever possible.

6.1.4 Composite Test Methods

Characterization of composite material systems requires that experimental tests be performed that yield *intrinsic* strength and stiffness properties of the material. Much effort has been expended in developing test methods that are appropriate for assessing material variations due to processing or material variables. However, these "quality assurance" test methods do not generally yield data that may be directly related to material performance. Although the test methods reviewed in this subsection were designed to provide intrinsic material property data, the degree of achievement of this goal varies from method to method. Therefore, as each test method is discussed in the following paragraphs, the nature of the material property data developed by a given test method will be discussed with respect to the desired intrinsic material property data.

The fundamental principle underlying these test methods can be stated as follows:

> The layer of given fiber orientation distribution is the building block of the laminate. Therefore, characterization of layer material properties allows predictions of the properties of any laminate.

In actual practice, the previous statement has been demonstrated to hold only for prediction of laminate elastic stiffness properties, such as effective modulus or Poisson's ratio. However, prediction of laminate strength properties from intrinsic layer (lamina) strength data has proven more difficult (see Volume 5). Therefore, in certain cases it is necessary to resort to direct characterization of laminate strength properties. With this necessity in mind, the following discussions will treat test methods for both layer and laminate characterizations.

Tensile Test Methods (ASTM D3039-76)

The tensile test is performed in order to determine uniaxial tensile strength, Young's modulus, and Pois-

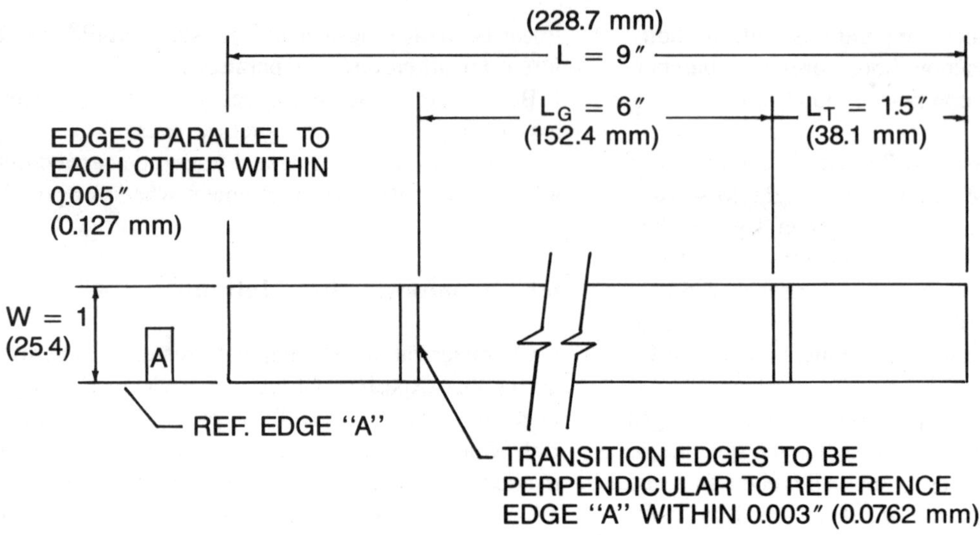

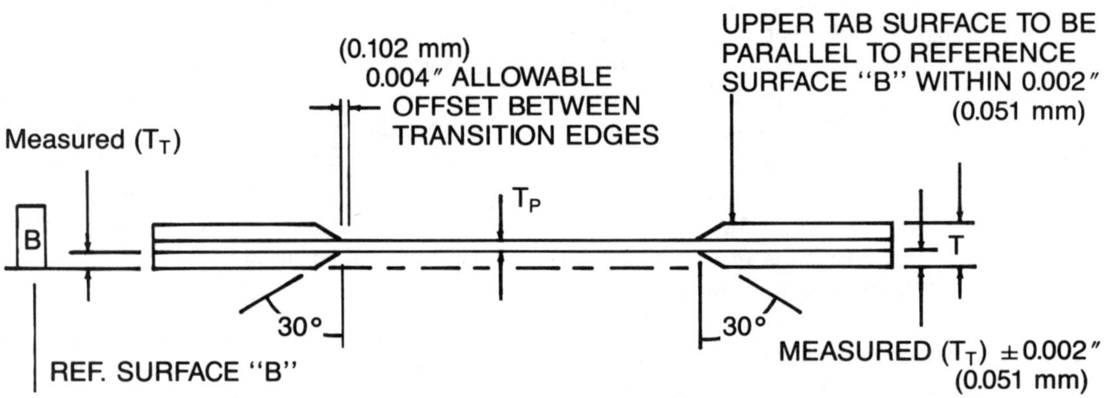

W	L	L_G	L_T	T	T_T	T_P	
±0.0156	±0.1250	±0.0625	±0.0625	±0.0150	±0.0050	——	inches
0.396	3.175	1.5875	1.5875	0.381	0.127	——	mm

FIGURE 6.1-16. Geometry of tensile coupon.

son's ratio [1,2]. The tensile test method allows determination of the following properties for the lamina:

E_1 = Young's modulus in the principal direction
E_2 = Young's modulus transverse to the principal direction
ν_{12} = Major Poisson's ratio
ν_{21} = Minor Poisson's ratio
X_1^T = Tensile strength in principal direction
X_2^T = Tensile strength transverse to principal direction
e_1^T = Ultimate strain in the principal direction
e_2^T = Ultimate strain transverse to the principal direction

The tensile specimen is straight-sided and of constant cross section with adhesively bonded and beveled tabs for load introduction. The lamina 0° test specimen is 12.7 mm (0.500 in) in width and six plies in thickness, whereas the lamina 90° test specimen is 25.4 mm (1.000 in) in width and eight plies in thickness. The overall length of the specimen is 229 mm (9.0 in), and the test section is 152 mm (6.0 in). The test specimen geometry is illustrated in Figure 6.1-16. Test specimen thickness variation should not exceed ±2 percent. Specimens should be precision machined from plates with tabs bonded in place. Test specimen edges should be undamaged and parallel to within 0.127 mm (5 mils). Tab surfaces should be parallel to a reference surface within 0.05 mm (2 mils).

The tensile test is performed utilizing wedge action friction grips. The specimen is first aligned in the grips and tightened in place. The specimen is then loaded monitonically to failure at a recommended rate of 0.02 cm/min. In order to determine specimen strains, electrical resistance strain gages are mounted on the specimen and monitored during the test. Foil gages with a resistance of 350 ohms and a gage length of 3.2 mm or 6.4 mm (0.125 and 0.250 in) have been found to minimize gage heating during the test. Typically, three gages are applied to each specimen: two longitudinal and one transverse to the load direction. The gages are located at the geometric center of the test section with the longitudinal gages mounted on opposite faces of the specimen. The gages may be monitored continuously or at discrete load intervals. If discrete data are taken, sufficient data points to established modulus must be taken within the region of linear elastic behavior. Typical test results for the 0° tensile test are shown in Figure 6.1-17 for an *E*-glass/epoxy composite. Test results for

the 90° tensile test of a boron/epoxy composite are shown in Figure 6.1-18. The relevant test method is ASTM D3039-76.

Tensile tests of laminate specimens of less than 2.5 mm (0.1 in) in thickness are best accomplished with a specimen geometry and test procedure identical to those discussed previously for the lamina 90° tensile tests. Such a test will allow determination of the following laminate "effective" properties:

E_x = effective laminate modulus in loading direction
ν_{xy} = effective laminate Poisson's ratio
$\overline{X}^T$ = laminate uniaxial ultimate stress
$\overline{e}^T$ = laminate uniaxial ultimate strain

Typical laminate tensile test results for the $[0_2/\pm45]_s$ boron/aluminum laminate are shown in Figure 6.1-19. Note that it is necessary to measure discrete stress-strain results at small stress intervals in order to establish the onset of the aluminum matrix yield.

Flexure Test Methods (ASTM 790-71)

The flexure test is based on the loading configuration shown in Figure 6.1-20, where the 3-point flexure test is shown in Figure 6.1-20(a) and the 4-point flexure test in Figure 6.1-20(b). It should be noted that the 4-point flexure test is sometimes utilized with the load points at $L/3$ as well as at quarter points as shown. The quarter-point loading has been utilized primarily for high modulus materials such as graphite/epoxy and boron/epoxy. These test methods are used for determining flexural strength and modulus. The test is not recommended for generating design data, but it does provide a simple test for quality control. If proper interpretation is given to the data, however, it may have more value than a quality control tool. This interpretation will be discussed in detail in the next segment.

The flexure methods are applicable to polymeric and composite materials that can be treated as homogeneous. For laminated materials the results will depend on ply stacking sequence, which requires the data to be interpreted in conjunction with laminated beam theory [19,20]. Because of this complexity, flexure tests in conjunction with high modulus, continuous filament composites are generally limited to unidirectional materials with fibers oriented at either 0° or 90° to the beam axis.

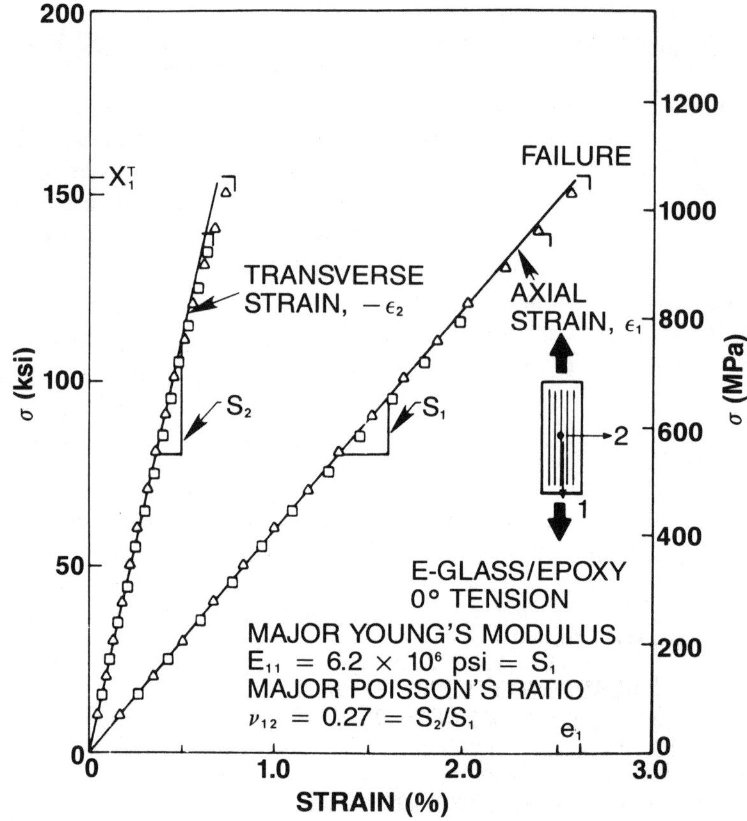

FIGURE 6.1-17. Unidirectional 0° tensile stress-strain behavior of *E*-glass/epoxy.

A testing machine with a controllable cross-head speed is used in conjunction with a loading fixture for a 4-point flexure test. Specific requirements for the radius of the load noses and supports are given in ASTM D790-71. Span-to-depth ratios, L/h, depend on the ratio of tensile strength parallel to the beam axis to interlaminar shear strength. For strength ratios less than 8 to 1, an L/h ratio of 16 is recommended. This ratio is typical of fiberglass composites. For high modulus materials such as graphite/epoxy and boron/epoxy, a value of $L/h = 32$ is recommended for 0° unidirectional composites. For 90° unidirectional properties of these materials, $L/h = 16$ is appropriate. Recommended specimen dimensions can also be found in ASTM D790-71. Cross-head travel rate, R, should be chosen such that the strain rate of the outer surface, $\dot{\epsilon}$, is approximately 0.01 mm/mm-min (0.01 in/in-min) for both the 3-point flexure and 4-point flexure at quarter points:

$$\dot{R} = \frac{\dot{\epsilon}L^2}{6h} \qquad (6.1\text{-}20)$$

For tensile strength determination the specimen is loaded until failure occurs. The maximum tensile stress, σ_m, is determined for the 3-point loading from the relationship

$$\sigma_m = \frac{3PL}{2bh^2} \qquad (6.1\text{-}21)$$

where b is the beam width. For 4-point loading at quarter points,

$$\sigma_m = \frac{3PL}{2bh^2} \qquad (6.1\text{-}22)$$

It should be noted that Equations (6.1-21) and (6.1-22) are valid only for materials in which the stress-strain curve is linear to failure. If some nonlinear stress-strain behavior occurs, a slight error will be introduced into the relationships for σ_m. An adjustment in the calculation for σ_m must also be made if the deflections become large (see ASTM D790-71).

For the determination of modulus, a deflectometer, an extensometer, or a linear voltage differential transformer (LVDT) can be used to measure center deflection. If a deflectometer is used, the test must be stopped and a reading taken at a point where the stress-strain behavior is still linear. The use of an extensometer or LVDT allows a continuous plot of load-deflection. Cross-head travel does not provide a very accurate measurement of deflection due to machine compliance. Although the error will be less for large strains, modulus is determined from the initial linear portion of the load-deflection curve where strains are usually small. Modulus is determined for the 3-point flexure test from the relationship

$$E_x = \frac{PL^3}{4bh^3w} (1 + S) \qquad (6.1\text{-}23)$$

where w is the center deflection and S is a correction

factor for shear deformation [20] given by

$$S = \frac{3h^2 E_x}{2L^2 G_{xz}} \qquad (6.1\text{-}24)$$

and G_{xz} is the shear modulus of the composite through the thickness. For 4-point loading at quarter points [20],

$$E_x = \frac{PL^3}{64bh^3w} (11 + 8S) \qquad (6.1\text{-}25)$$

Note that the shear correction factor is a function of L/h and E_x/G_{xz}. For high modulus materials such as graphite/epoxy and boron/epoxy, E_x/G_{xz} is greater than 20 for 0° unidirectional material. Thus, shear deformation will be significant even for a span-to-depth ratio of 32. In many cases, G_{xz} may not be known, but an approximate value of E_x/G_{xz} is known. If a value of G_{xz}

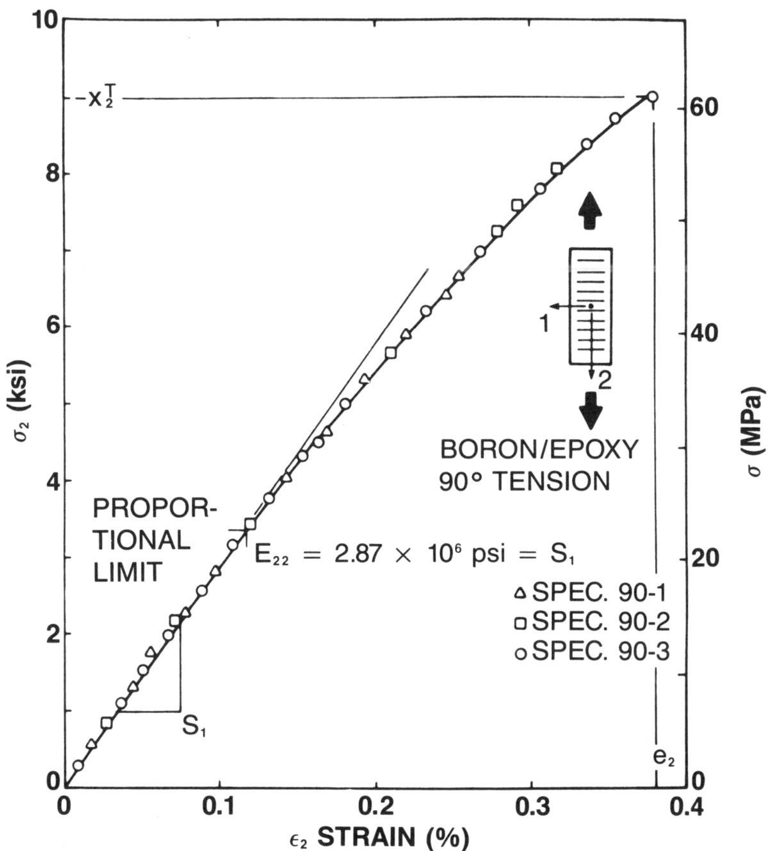

FIGURE 6.1-18. Unidirectional 90° tensile stress-strain behavior of boron/epoxy.

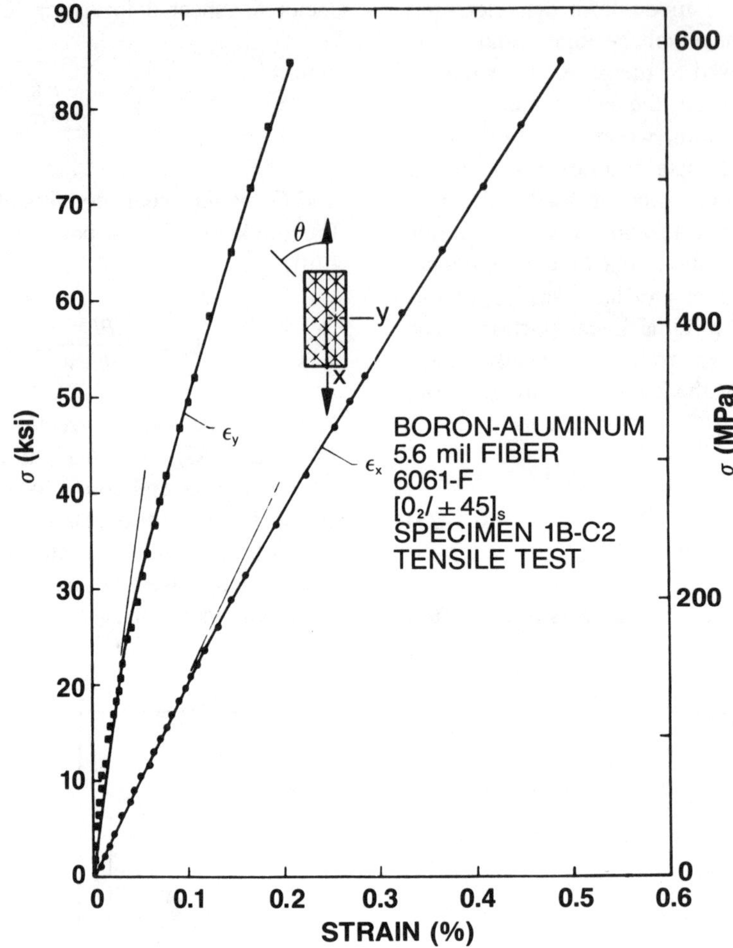

FIGURE 6.1-19. Tensile stress-strain behavior of boron/aluminum laminate.

can be determined, Equations (6.1-23) and (6.1-25) can be used in the form

$$E_x = \frac{11PL^3}{4bh^3\left[w - \dfrac{3PL}{8bhG_{xz}} \right]} \qquad (6.1\text{-}26)$$

for 4-point loading at quarter points. If shear deformation is neglected, Equations (6.1-23) and (6.1-25) take the simplified form

$$E_x = \frac{PL^3}{4bh^3w} \qquad (6.1\text{-}27)$$

for 3-point loading, and

$$E_x = \frac{11PL^3}{64bh^3w} \qquad (6.1\text{-}28)$$

for 4-point loading at quarter points.

A cursory examination of Equations (6.1-23) and (6.1-25) reveals that the neglect of shear deformation yields an apparent modulus that is less than the actual value of E_x. Thus, if the shear correction factor, S, cannot be easily determined, the flexure tests can be run for modulus at increasing ratios of L/h until a constant value of E_x is measured. For high values of L/h, however, large deflections may occur at low stress levels. Thus, the applied load should be just sufficient to provide an initial slope of the load-deflection curve. Obviously, it is also necessary to keep loads below specimen damage levels.

The 4-point load test is often preferred to the 3-point method because the center section is under pure bending stresses. Shear stresses do exist, however, between the outer supports and the applied load. The maximum shear stress, τ_m, for 3-point loading is given by

$$\tau_m = \frac{h}{2L}\,\sigma_m \qquad (6.1\text{-}29)$$

and for 4-point loading at quarter points, τ_m is given by

$$\tau_m = \frac{h}{L}\,\sigma_m \qquad (6.1\text{-}30)$$

Thus, for fixed values of L/h, the maximum shear stress in the 4-point test is twice the maximum shear stress in the 3-point loading. Equations (6.1-29) and (6.1-30) can be used to determine an L/h ratio that will assure tensile failure. When flexure tests are run at elevated temperatures, the shear strength may drop significantly, causing shear failures. As a result, lower span-to-depth ratios may be necessary to produce elevated temperature tensile failures.

Comparison between Tensile and Flexure Strength

Tensile data generated from a unidirectional flexure test usually yield higher strength than data obtained from a standard tensile coupon. It is primarily for this reason that flexure data are not considered appropriate for design purposes. This difference in apparent tensile strengths can be partly accounted for, however, if one considers the brittle nature of most polymeric matrix composites. In particular, the Weibull statistical

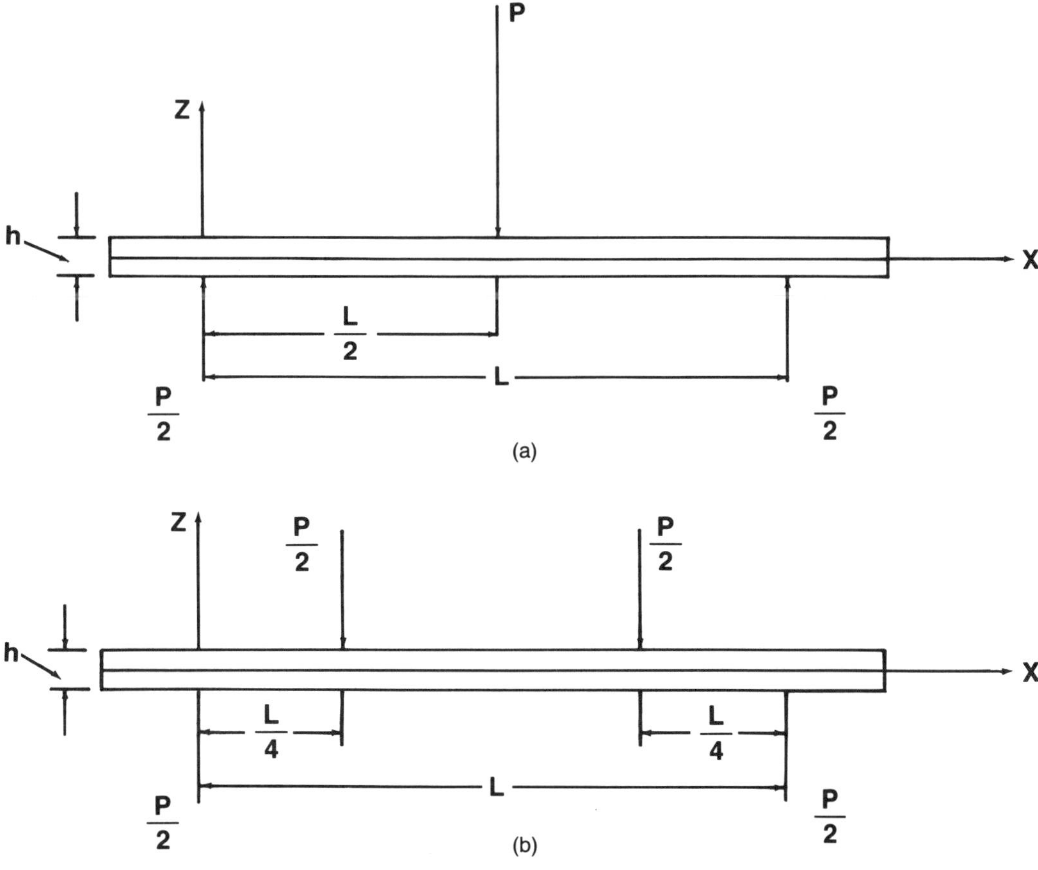

FIGURE 6.1-20. Flexure test: (a) 3-point; (b) 4-point.

strength theory for brittle materials [21] states that the probability of survival, P, at a stress level, σ, for a uniaxial stress field in a homogeneous material governed by a volumetric flaw distribution is given by

$$P(\sigma) = \exp\left[-B(\sigma)\right] \qquad (6.1\text{-}31)$$

where B is the risk of rupture. For a two-parameter Weibull model that has been used in conjunction with composite materials [22–24], the risk of rupture is given by the volume integral:

$$B(\sigma) = \int_v \left(\frac{\sigma}{\sigma}\right)^a dV \qquad (6.1\text{-}32)$$

where σ is a location parameter of the distribution, often referred to as the *characteristic strength*, and a is the shape parameter. Both of these parameters are considered to be material properties independent of size. Thus, the risk to break will be a function of the stress distribution in the test specimen. For the case of a simple tensile test under uniform stress, Equation (6.1-32) becomes

$$B_t = V_t \left(\frac{\sigma_t}{\sigma}\right)^a \qquad (6.1\text{-}33)$$

where the subscript t denotes simple tension. Equation (6.1-31) can now be written in the form

$$P(\sigma_t) = \exp\left[-\left(\frac{\sigma_t}{\sigma_t}\right)^a\right] \qquad (6.1\text{-}34)$$

where σ_t is the location parameter for tensile loading given by

$$\sigma_t = \frac{\sigma}{(V_t)^{1/a}} \qquad (6.1\text{-}35)$$

Thus, tensile tests from specimens of different dimensions could be represented by a two-parameter Weibull distribution with the same shape parameter, according to Equation (6.1-35).

The integration in Equation (6.1-32) has been performed by Weil and Daniel [25] for the 3-point flexure specimens:

$$B_{3f} = \frac{V_{3f}}{2(a+1)^2}\left(\frac{\sigma_{3f}}{\sigma}\right)^a \qquad (6.1\text{-}36)$$

and for the 4-point flexure test at quarter points:

$$B_{4f} = \frac{V_{4f}(a+2)}{4(a+1)^2}\left(\frac{\sigma_{4f}}{\sigma}\right)^a \qquad (6.1\text{-}37)$$

where the subscripts $3f$ and $4f$ denote 3-point flexure and 4-point flexure, respectively. The stresses σ_{3f} and σ_{4f} are the stresses at the outer surface of the flexure specimen.

Equations (6.1-36) and (6.1-37) in conjunction with Equation (6.1-31) yield

$$P(\sigma_{3f}) = \exp\left[-\left(\frac{\sigma_{3f}}{\sigma_{3f}}\right)^a\right] \qquad (6.1\text{-}38)$$

$$P(\sigma_{4f}) = \exp\left[-\left(\frac{\sigma_{4f}}{\sigma_{4f}}\right)^a\right] \qquad (6.1\text{-}39)$$

where

$$\sigma_{3f} = \sigma\left[\frac{2(a+1)^2}{V_{3f}}\right]^{1/a} \qquad (6.1\text{-}40)$$

$$\sigma_{4f} = \sigma\left[\frac{4(a+1)^2}{(a+2)V_{4f}}\right]^{1/a} \qquad (6.1\text{-}41)$$

Thus, both the 3-point flexure and 4-point flexure tests yield two-parameter Weibull distributions with the same shape factor as the tensile test but with different characteristic strengths.

Using Equations (6.1-34), (6.1-40), and (6.1-41) yields the following ratios for characteristic flexure strength to tensile strength:

$$\frac{\sigma_{3f}}{\sigma_t} = \left[2(a+1)^2\left(\frac{V_t}{V_{3f}}\right)\right]^{1/a} \qquad (6.1\text{-}42)$$

$$\frac{\sigma_{4f}}{\sigma_t} = \left[\frac{4(a+1)^2}{(a+2)}\left(\frac{V_t}{V_{4f}}\right)\right]^{1/2} \qquad (6.1\text{-}43)$$

In order to illustrate the effect of nonuniform stress distribution, consider the case $V_t = V_{3f} = V_{4f}$. For values of $a = 15$ and 25, Equations (6.1-42) and (6.1-43) yield

$$\frac{\sigma_{3f}}{\sigma_t} = \begin{cases} 1.52, & a = 15 \\ 1.33, & a = 25 \end{cases} \qquad (6.1\text{-}44a)$$

$$\frac{\sigma_{4f}}{\sigma_t} = \begin{array}{l} 1.31, \ a = 15 \\[6pt] 1.20, \ a = 25 \end{array} \qquad (6.1\text{-}44b)$$

These values of a are typical of currently utilized composites such as glass/epoxy and graphite/epoxy. Thus, the flexure test can produce significantly higher tensile strengths than the tensile test, with the 3-point loading producing the highest strength. This result is due to the fact that the maximum stress is produced at the outer surface in the center of the beam, while the 4-point loading produces the maximum stress at the outer surface throughout the center section. In particular, the smaller the volume under maximum stress, the higher the local strength.

Off-Axis Tensile Test

The tensile test method is based on ASTM Standard D3039-76, which is applicable to composites with orthotropic in-plane properties. It is often desirable, however, to measure unidirectional tensile properties at some angle relative to the principal material coordinate system, which is accomplished by utilizing an off-axis tensile test. The off-axis test specimen consists of a laminate with all layers at equal fiber orientation, θ, with respect to the longitudinal axis of the specimen. This test method is often used to measure off-axis tensile modulus for the purpose of checking the validity of compliance transformation equations. It also provides a method of applying biaxial stress relative to the principal material coordinate system.

For the case of uniaxial tension, σ_x, in conjunction with the engineering constants, the strain components are

$$\epsilon_x = \frac{1}{E_x}\sigma_x \qquad (6.1\text{-}45)$$

$$\epsilon_y = \frac{-\nu_{xy}}{E_x}\sigma_x \qquad (6.1\text{-}46)$$

$$\gamma_{xy} = \frac{\eta_{xy}}{E_x}\sigma_x \qquad (6.1\text{-}47)$$

Thus, the material is anisotropic relative to the load direction, and shear coupling is present. It is the presence of the shear coupling phenomenon that causes ad-

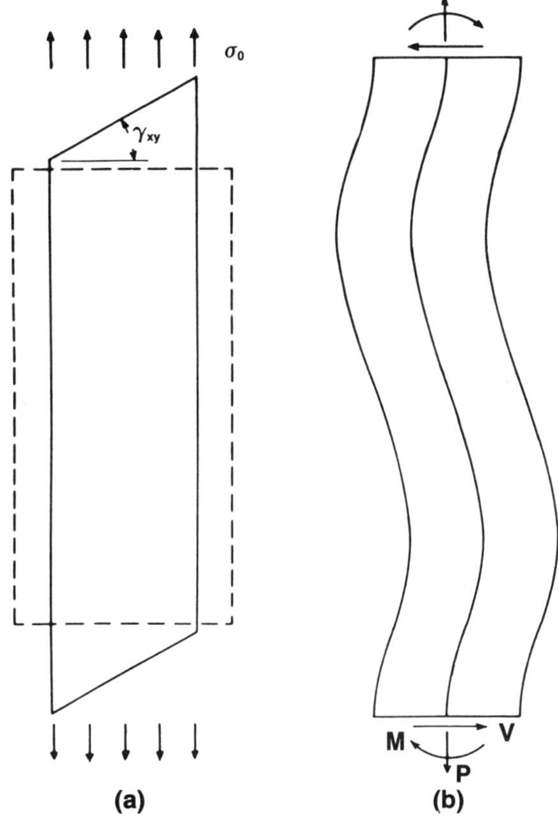

FIGURE 6.1-21. Effect of end constraint on off-axis specimen: (a) uniform state of stress; (b) effect of clamped ends.

ditional complications in the off-axis tensile test compared to the tensile method for orthotropic composites covered by ASTM D3039-76. In particular, the clamping of the tensile coupon at the ends prohibits local rotation of the specimen, which induces a nonuniform strain field as illustrated in Figure 6.1-21. It has been shown by Pagano and Halpin [26], however, that a uniform state of stress and strain will exist at the center of an off-axis tensile coupon if the length-to-width ratio, L/W, is sufficiently large. The exact dimensions necessary to accomplish this depend on the degree of shear coupling as quantified by the shear coupling ratio η_{xy} [1,2]. If an apparent modulus, E_{x*}, is experimentally determined in the usual manner for an off-axis tensile coupon, the error introduced by shear coupling can be estimated:

$$E_x = (1 - \eta)E_{x*} \qquad (6.1\text{-}48)$$

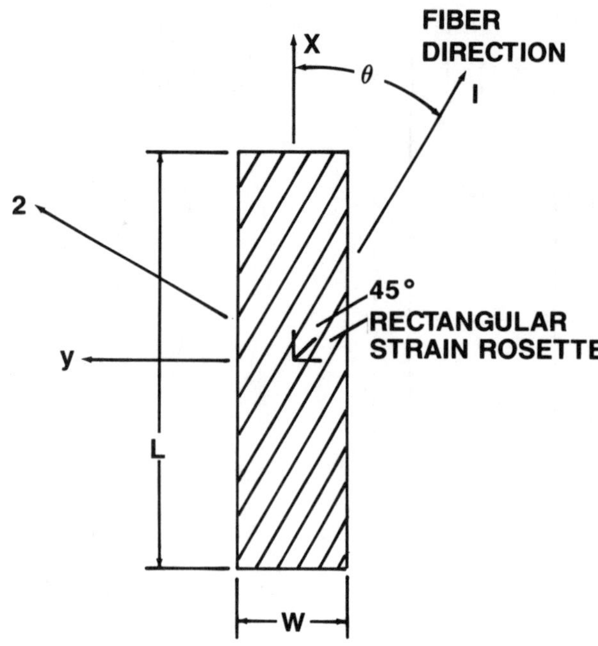

FIGURE 6.1-22. Off-axis tensile coupon. $\theta < 0$.

where

$$\eta = \frac{3\eta_{xy}}{\left(\dfrac{3E_x}{G_{xy}} + \dfrac{2L^2}{W^2} \right)} \qquad (6.1\text{-}49)$$

Thus, η is a direct measure of the error involved in the observed modulus. A cursory examination of Equation (6.1-49) reveals that η vanishes for decreasing values of η_{xy} or increasing ratios of L/W. The test procedure for the off-axis tensile test is essentially the same as described in D3039-76. Modifications of this specimen have been proposed by Cron et al. [27].

If the off-axis tensile test is used as a means of introducing biaxial strain relative to the principal material coordinate system, a three-element (rosette) gage is necessary to determine the stress-strain response parallel to the fibers, transverse to the fibers, and in longitudinal shear. The data reduction procedure consists of recording the axial load and the output of the three strain gages. For the case of uniaxial tension ($\sigma_y = \tau_{xy} = 0$),

$$\sigma_1 = m^2 \sigma_x$$

$$\sigma_2 = n^2 \sigma_x \qquad (6.1\text{-}50)$$

$$\tau_{12} = -mn\sigma_x$$

where $m = \cos \theta$, $n = \sin \theta$, and θ is defined in Figure 6.1-22. Using a rectangular rosette gage, as shown in Figure 6.1-22, the strain at 45° can be determined with the result,

$$\epsilon_{45} = \frac{1}{2} \epsilon_x + \frac{1}{2} \epsilon_y + \frac{1}{2} \gamma_{xy} \qquad (6.1\text{-}51)$$

where ϵ_{45} denotes the normal strain at $\theta = 45°$. Solving Equation (6.1-51) for γ_{xy} yields

$$\gamma_{xy} = 2\epsilon_{xy} - \epsilon_x - \epsilon_y \qquad (6.1\text{-}52)$$

Substituting Equation (6.1-52) into the transformation equation yields

$$\epsilon_1 = m(m - n)\epsilon_x + n(n - m)\epsilon_y + 2mn\epsilon_{45} \qquad (6.1\text{-}53)$$

$$\epsilon_2 = n(m + n)\epsilon_x + n(n + m)\epsilon_y + 2mn\epsilon_{45} \qquad (6.1\text{-}54)$$

$$\gamma_{12} = -(m^2 + 2mn - n^2)\epsilon_x - (m^2 - 2mn - n^2)\epsilon_y$$
$$+ 2(m^2 - n^2)\epsilon_{45} \qquad (6.1\text{-}55)$$

Thus, the strains relative to the fiber orientation can be expressed in terms of the strain rosette measurements ϵ_x, ϵ_y, and ϵ_{45}. Equations (6.1-45) and (6.1-53) through (6.1-55) allow the stress-strain curves parallel to the fibers, transverse to the fibers, and in longitudinal shear to be determined under biaxial load. As will be discussed in a later paragraph, the off-axis tensile coupon can be used as a means of measuring unidirectional shear response.

It should be noted that other three-element strain gage configurations can be used. The rectangular configuration, however, provides the simplest formulas for data reduction.

In summary, the off-axis tensile test is run essentially the same as the tensile test for on-axis composites with the exception of the choice of specimen dimensions that must be designed to minimize shear coupling effects [28] and the use of a three-element strain gage if complete stress-strain behavior relative to the fiber orientation is desired. Equations (6.1-49) and (6.1-50) can be utilized in determining specimen dimensions. A rectangular rosette gage, in conjunction with Equations (6.1-53) through (6.1-55), provides a straightforward means of determining strain response relative to the fiber orientation.

Typical off-axis stress-strain response is shown in Figure 6.1-23 for a 45° boron/epoxy composite. A two-element rectangular strain gage was used to obtain response in the x and y directions.

Compression Test Methods (ASTM 3410-75)

Perhaps the most difficult of the intrinsic material properties of composites to measure are the compressive strength properties. This is because slight specimen geometric variations result in eccentricity of the applied load, thereby enhancing the opportunity for failure to occur due to geometric instability. Thus, in order to achieve an accurate measure of the compressive strength of a given composite material, rather complex loading fixtures and specimen configurations have been developed [1].

In this segment three generic compression test methods will be discussed. The first method (Type I) is characterized by a specimen with a short test section length that is completely unsupported. The test specimen is generally loaded through friction by means of wedge action friction grips. Associated test fixtures, which insure colinearity of the applied load and specimen centerline, are required. To meet this need, several test fixture configurations have been developed. One such fixture, shown in Figure 6.1-24, is the Celanese test fixture (ASTM D-3410-75). The Celanese fixture employs truncated conical friction grips contained in matching cylindrical end fittings. Colinearity of the cylindrical end fittings is insured by a hollow cylinder that contains the fittings. The test specimen geometry is also shown in Figure 6.1-24, where the specimen is shown to be 141 mm (5.5 in) in length and 6.4 mm (0.25 in) in width with a test section of 12.7 mm (0.5 in). Beveled end tabs are bonded to the specimen for load introduction. Strain gages can be used with this specimen to measure modulus.

A second example of the Type I compression test method is the IITRI test method developed by Illinois Institute of Technology Research Institute. The IITRI test fixture employs a test specimen identical in geometry to that of the Celanese test method. Shown in Figure 6.1-25, the IITRI test method employs linear bearings and hardened shafts to insure colinearity of the load path. Load is applied to the specimen through serrated wedges that are contained in solid steel bases. The IITRI fixture may also be modified to accept specimens of widths greater than 6.4 mm (0.25 in). In addition,

support rollers to restrain column instability of the specimen may be added to the wedge action friction grips. These grip modifications are shown in Figure 6.1-26.

Another example of the Type I compression test method is the Northrop method. The Northrop compression test utilizes offset unsupported lengths, as shown in Figure 6.1-27. Stability is provided by the thick side supports. This method offers the advantage of a much simpler test fixture compared to the Celanese and ITTRI methods.

The final example of the Type I compression test method was developed by the National Institute of Standards and Technology (NIST). The NIST test fixture combines certain features of the IITRI and Celanese test fixtures while introducing a feature that allows tensile loading. As shown in Figure 6.1-28, the NIST compression fixture consists of a test specimen contained in end fixtures that are constrained to move in a colinear fashion by rigid rods and an external housing. Specimen gripping is achieved by friction due to interference between end fixtures and cylindrical epoxy specimen building. Overall, specimen length is 11.8 mm (4.65 in) while the test section is 16 mm (0.63 in). This method utilizes both square cross-sectional and round cross-sectional specimens. The round cross section is recommended for zero degree unidirectional composites only.

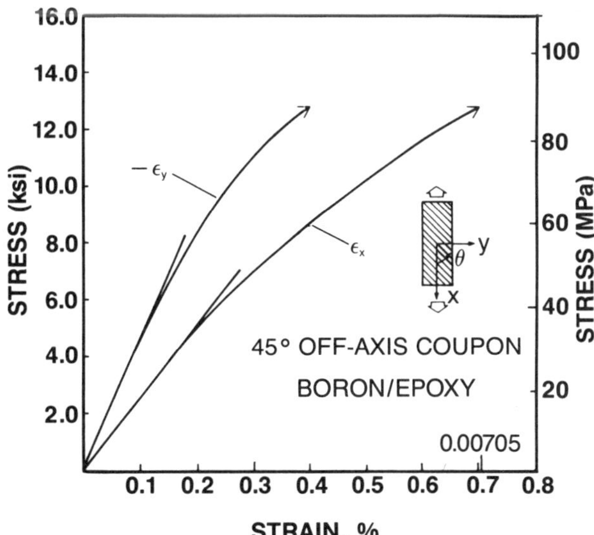

FIGURE 6.1-23. Tensile stress-strain response of 45° off-axis coupon.

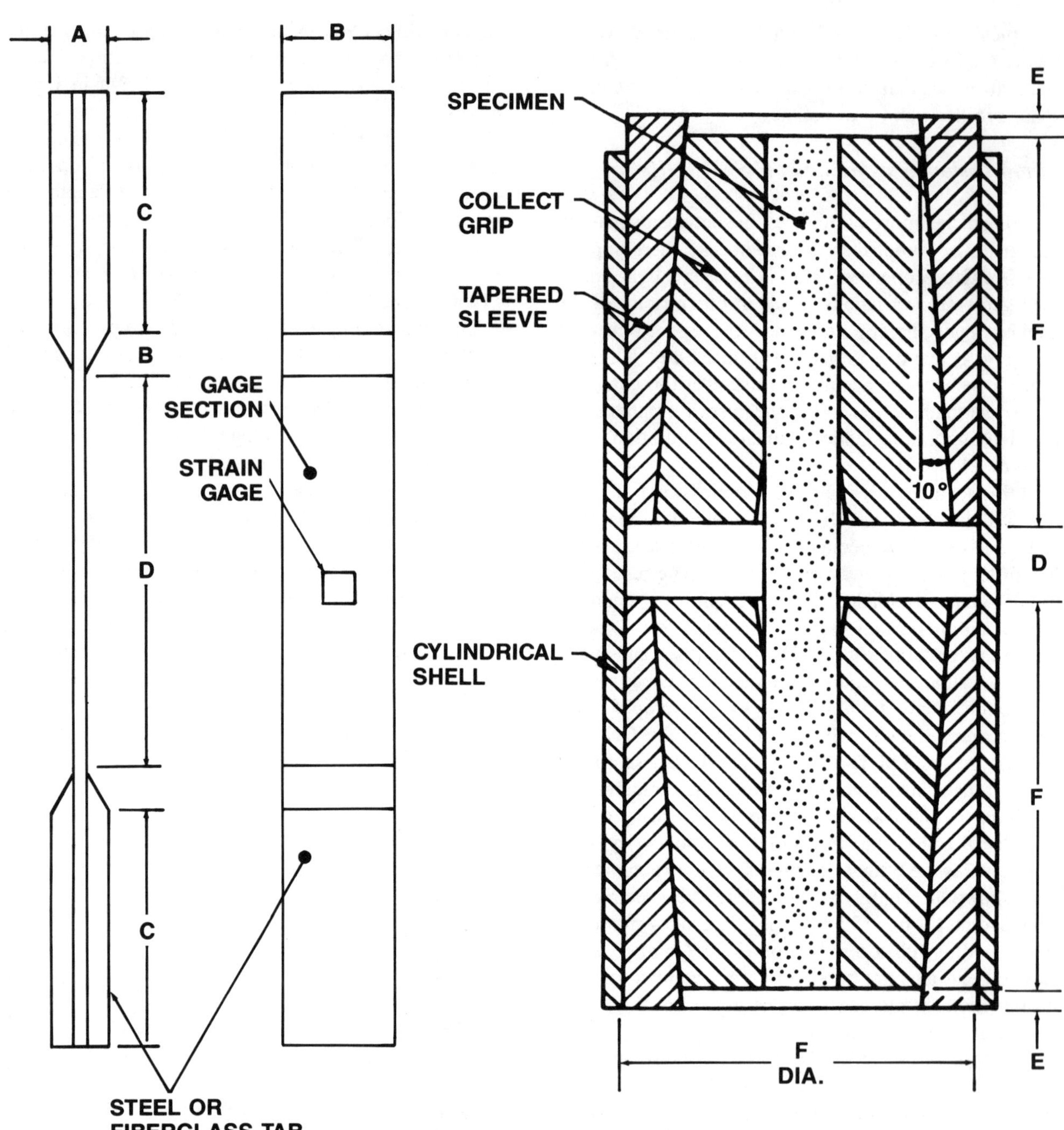

DIMENSION	mm	inches
A	39.9	0.157
B	6.35	0.250
C	57.2	2.250

DIMENSION	mm	inches
D	12.7	0.500
E	3.18	0.125
F	63.5	2.500

FIGURE 6.1-24. Celanese compression specimen and test fixture, ASTM D3410-75.

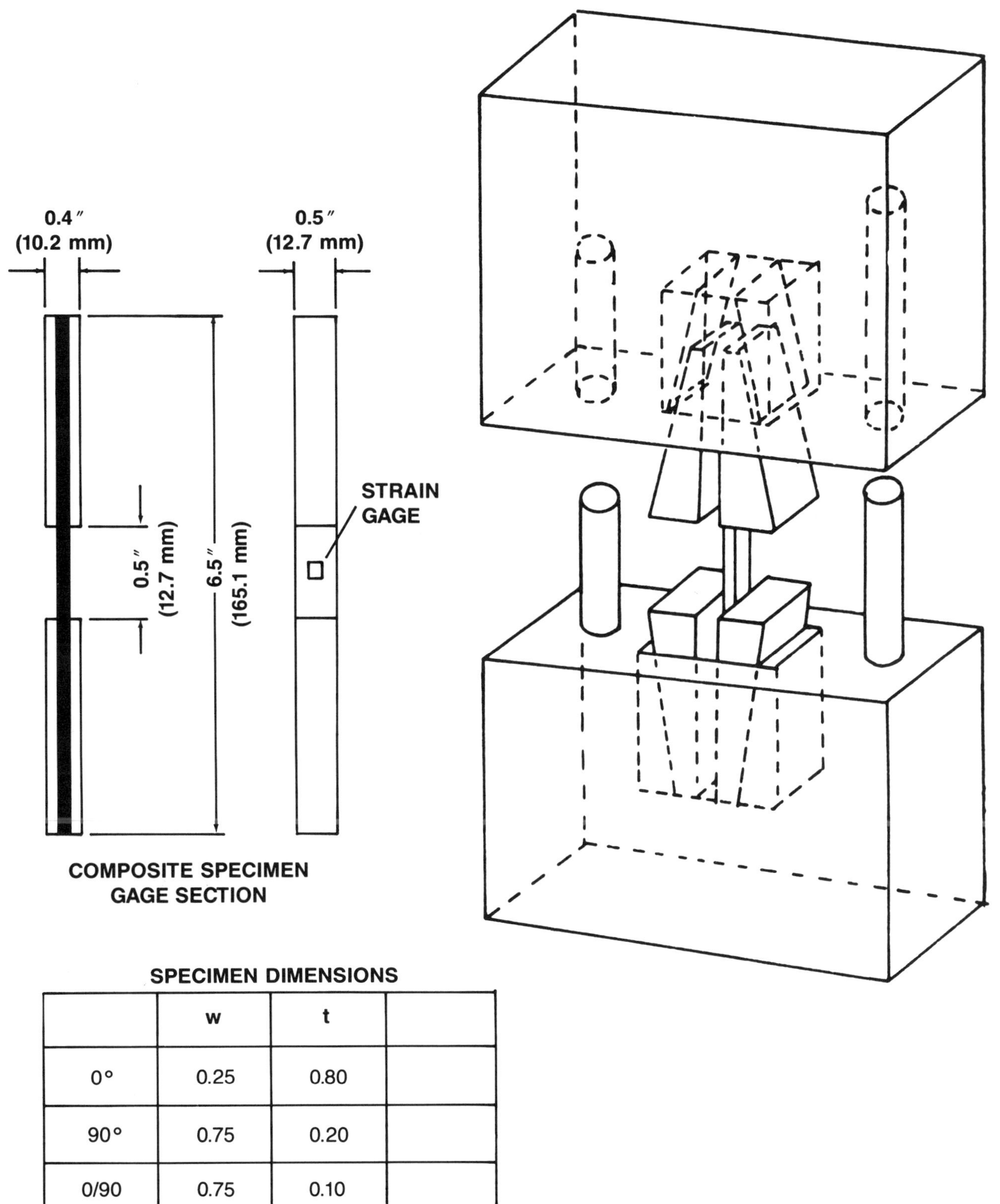

SPECIMEN DIMENSIONS

	w	t	
0°	0.25	0.80	
90°	0.75	0.20	
0/90	0.75	0.10	

FIGURE 6.1-25. IITRI compression test fixture and specimen.

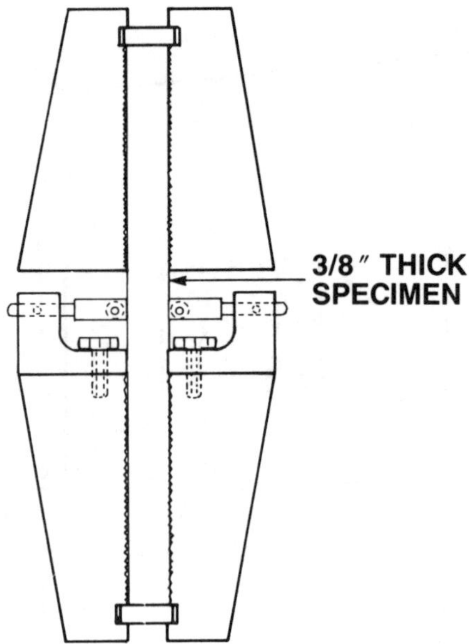

FIGURE 6.1-26. Modified grips for IITRI compression fixture.

The four Type I compression test methods appear to yield acceptable data. However, certain problems may be encountered in their implementation. The Celanese test method requires extreme precision in mounting the specimen in the fixture. The mass of the IITRI test fixture requires prolonged soak periods for elevated temperature tests. All of the Type I test fixtures require that test specimens be fabricated such that specimen edges and/or tab surfaces are extremely parallel. Since the tabs are generally adhesively bonded to the specimen surfaces, bonding fixtures are required to insure the desired parallelism. In addition, composite specimens fabricated with vacuum bag techniques may exhibit rather large thickness variations. Compressive tests of specimens prepared for these laminates may lead to erroneous results unless great care is taken to align the test specimen.

The second class of compression test methods (Type II) is characterized by a specimen of relatively long test section that is fully supported. The first example of the Type II compression test is the SWRI method developed by Southwest Research Institute [29]. The test specimen and support fixture are illustrated in Figures 6.1-29a and 6.1-29b. The SWRI test specimen geometry is a modification of the laminate tensile test specimen discussed earlier in which the load introduction tabs

have been lengthened from 38 mm (1.5 in) to 64 mm (2.5 in), and the specimen length has been reduced from 230 mm (9 in) to 218 mm (8 in). The test fixture provides contact support for the specimen over the entire test section length. A notch is cut in one support to allow a transverse strain gage to be placed on the specimen for the purpose of measuring Poisson's ratio in compression. Longitudinal strain is determined by using a strain gage or extensometer placed on the edge of the specimen. For laminates displaying a significant free edge effect (see Volume 5), such longitudinal strain measurements may be erroneous.

A second example of the Type II compression test method is the side supported fixture developed by Lockheed-California Company [30] illustrated in Figure 6.1-30. The Lockheed method utilizes side supports over the gage section of the specimen only. This is the major difference between this method and the SWRI method. In the Lockheed method the bottom end tab is cut off and the upper end tab, which extends outside the supports, is gripped. Thus, the load is transferred through shear rather than direct compression. It should be noted that the bottom end tab does not have to be cut off. The specimen would then be gripped at both top and bottom.

It is often desirable to determine the compression strength utilizing a tensile coupon. Such is the case when residual compression strength is to be determined after preloading in tension [29] or after subjecting a tensile coupon to fatigue loading. These requirements are a major reason for developing the Type II compression methods.

The Type II compression tests appear to yield laminate data that are compatible with Type I compression data. The values on zero degree unidirectional compression generated with fully supported coupons, however, appear to be consistently lower than the Type I compression data. There is no precise explanation for this behavior. Perhaps the increased stiffness of the zero degree unidirectional specimen makes alignment more critical in the case of the Type II compression method.

The third class of compression test method (Type III) involves the loading of a straight-sided coupon bonded to a honeycomb core that supplies the support necessary for stability. The first example of the Type III compression test is the sandwich edgewise compression test that utilizes two coupons bonded to a honeycomb core, as shown in Figure 6.1-31. The load is usually applied through self-aligning bearing blocks. A potting compound can be used to reinforce the ends, and straps can be clamped along the bottom of the flat laminate to provide a firm support for the loading heads and to prevent premature end crushing. Stresses are calculated assuming the core does not carry any load:

$$\bar{\sigma}_y = \frac{N_y}{h} = \frac{P}{2bh} \tag{6.1-56}$$

where $\bar{\sigma}_y$ is the average axial stress in the laminate. Strain gages mounted at the center of the face sheets can be used to measure E_y and ν_{yx} with the result

$$E_y = \frac{P}{2bh\epsilon_y} \tag{6.1-57}$$

$$\nu_{yx} = \frac{-\epsilon_x}{\epsilon_y} \tag{6.1-58}$$

where ϵ_x and ϵ_y are the strains measured by the strain gages. The initial linear portion of the load deformation curves is used to determine E_y and ν_{yx}. If the specimen is properly aligned, the opposite face strains should not vary more than 10 percent.

The second example of the Type III test method involves the loading of a honeycomb sandwich beam in 4-point bending with the composite face sheet on the compression side of the beam (see Figure 6.1-32). A metal face sheet is used on the tension side of the beam. Core and face sheet materials and beam dimensions are chosen such that failure occurs in the composite face sheet. Design parameters include metal face sheet strength, core shear strength, weight and cell size, strength of the adhesive and cure temperature, beam span and allowable mid-span deflection, and beam overhang and test section length. Some of these parameters depend on the specific composite material and laminate orientation, as discussed by Lantz [31]. The overall beam dimensions are usually constant, however, with a total length, including overhang, of 563 mm (22 in), a honeycomb core thickness of 38 mm (1.5 in), and a face sheet width of 25.5 mm (1 in).

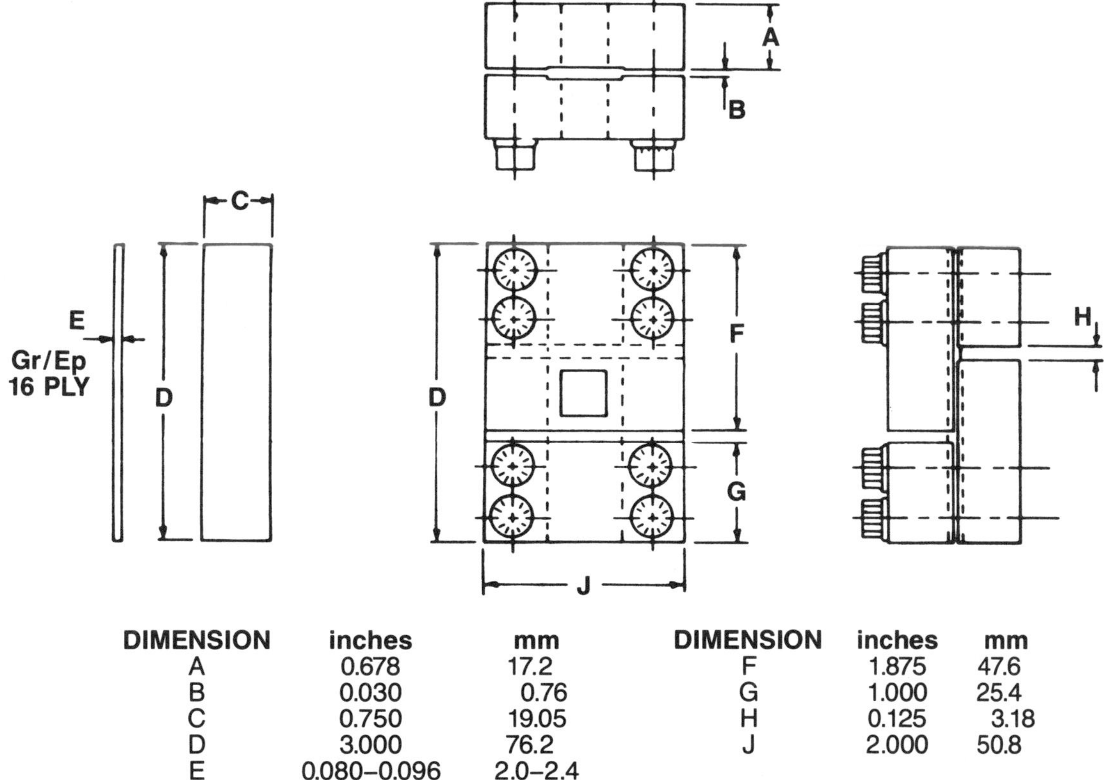

DIMENSION	inches	mm		DIMENSION	inches	mm
A	0.678	17.2		F	1.875	47.6
B	0.030	0.76		G	1.000	25.4
C	0.750	19.05		H	0.125	3.18
D	3.000	76.2		J	2.000	50.8
E	0.080–0.096	2.0–2.4				

FIGURE 6.1-27. Northrop compression specimen and test fixture.

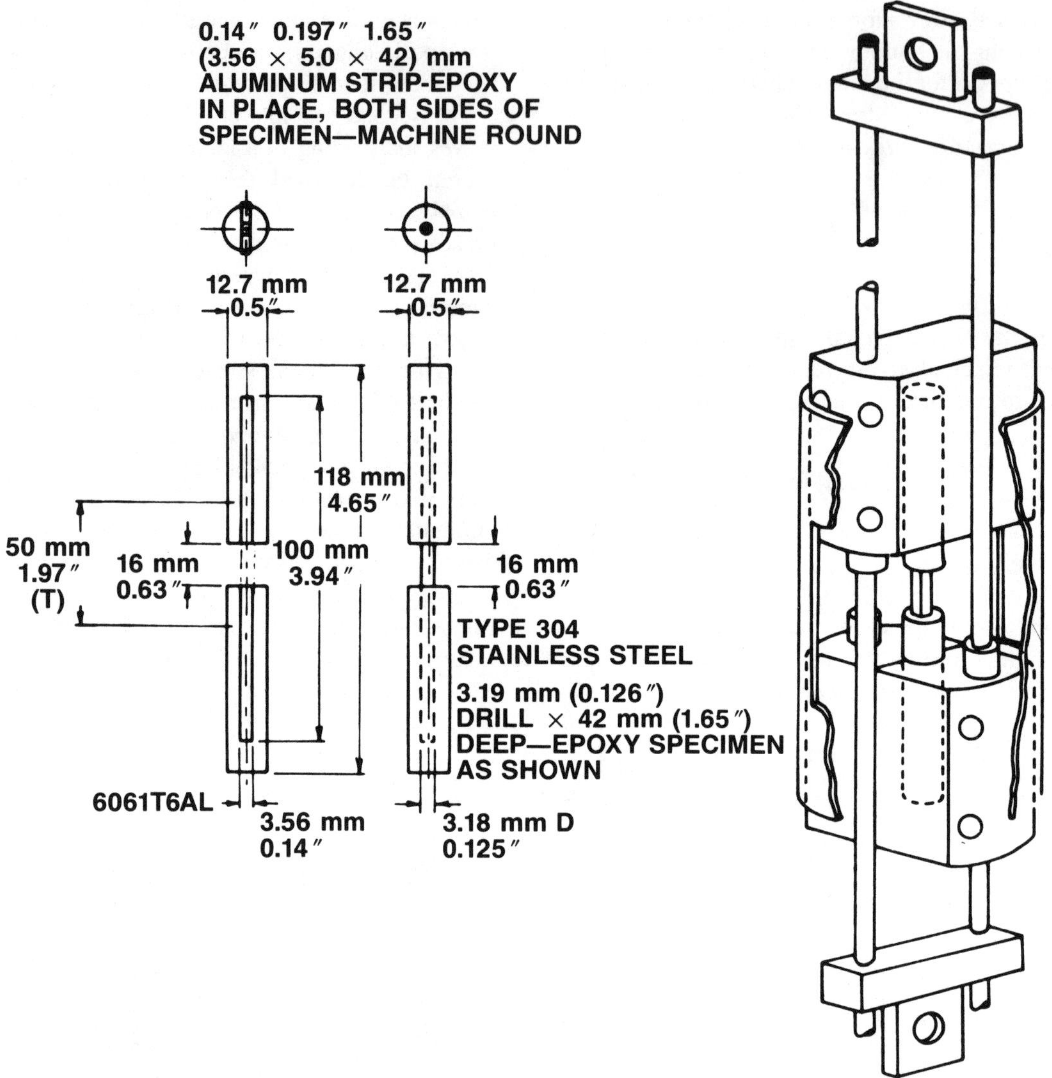

FIGURE 6.1-28. NBS compression test fixture and specimen.

As illustrated in Figure 6.1-32, strain gages can be bonded at the center of the specimen to measure E_x and ν_{xy}. Data obtained from sandwich beam specimens for determination of ν_{xy} have shown values that are often higher than measurements obtained from tensile coupons. Boundary effects or the presence of transverse curvature are possible sources of Poisson's ratio error in the sandwich beam specimen.

Two methods have been utilized in transferring load in the sandwich beam test. The first method involves direct load application to the composite face sheet in a manner similar to the 4-point flex test discussed in sub-section 6.1.4. If this method is used, hard rubber pads should be placed under the load noses to prevent local crushing of the composite face sheet. The second method uses load blocks buried in the core.

Laminate stress is determined by assuming uniform deformation in the face sheets and neglecting bending stresses in the core, with the result

$$\bar{\sigma}_x = \frac{N_x}{h} = \frac{PL}{4bh(2t + h + h')} \tag{6.1-59}$$

$$\nu_{xy} = \frac{-\epsilon_y}{\epsilon_x} \tag{6.1-60}$$

where ϵ_x and ϵ_y are the strains measured at the center section of the composite face sheet. Again, the initial linear portion of the load deformation curves is used to determine E_x and ν_{xy}.

It should be noted that the sandwich beam can also be used to determine tensile properties. Due to the success of the tensile coupon and the expensive nature of the sandwich beam specimen, the method is not widely utilized for tensile properties.

The sandwich beam specimen usually yields higher compression strength than any of the other methods.

Shear Test Methods (ASTM D3518-76, D2344-76)

Shear tests are performed in order to establish the shear strength, ultimate shear strain, and shear modulus of the composite material. Proper characterization of the shear properties of the composite lamina or laminates requires evaluation of three distinct moduli and strengths. The lamina properties in the plane of lamination (1–2) are known as *in-plane shear properties*, whereas the properties in the 1–3 and 2–3 plane are

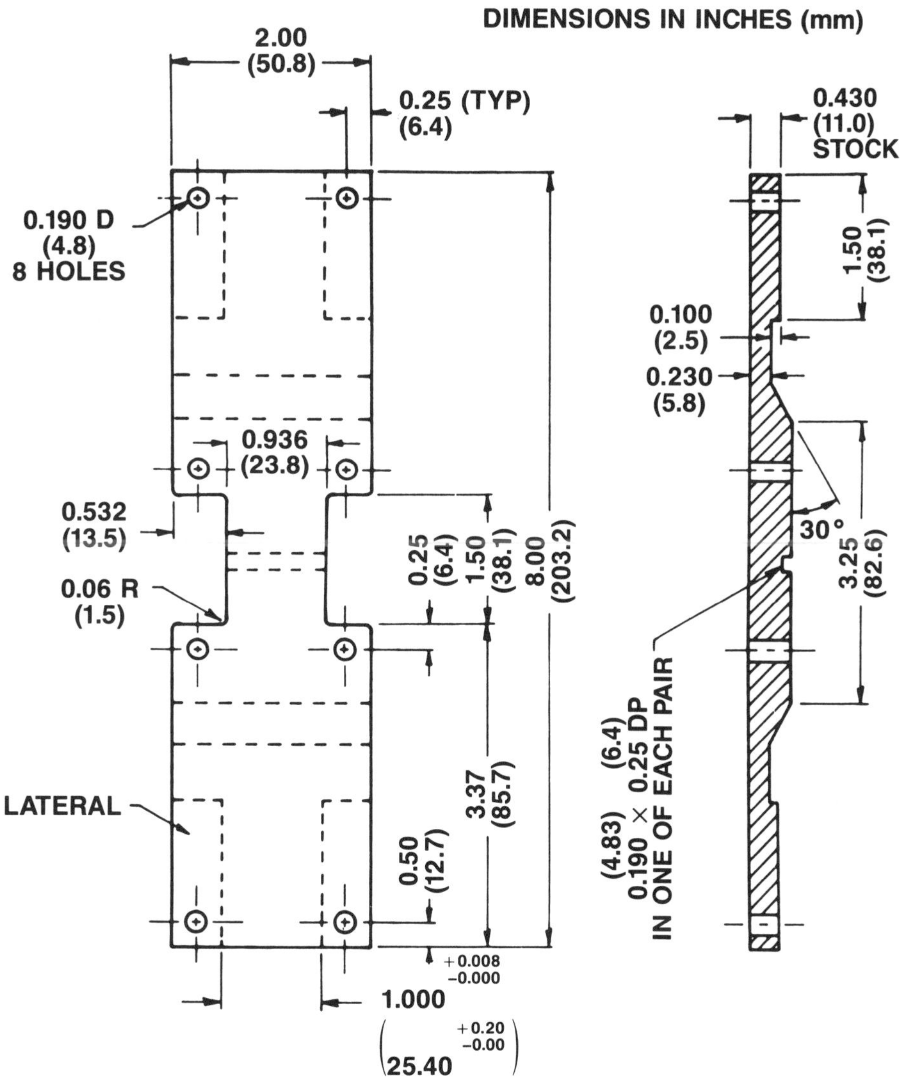

FIGURE 6.1-29a. SWRI fully supported compression fixture.

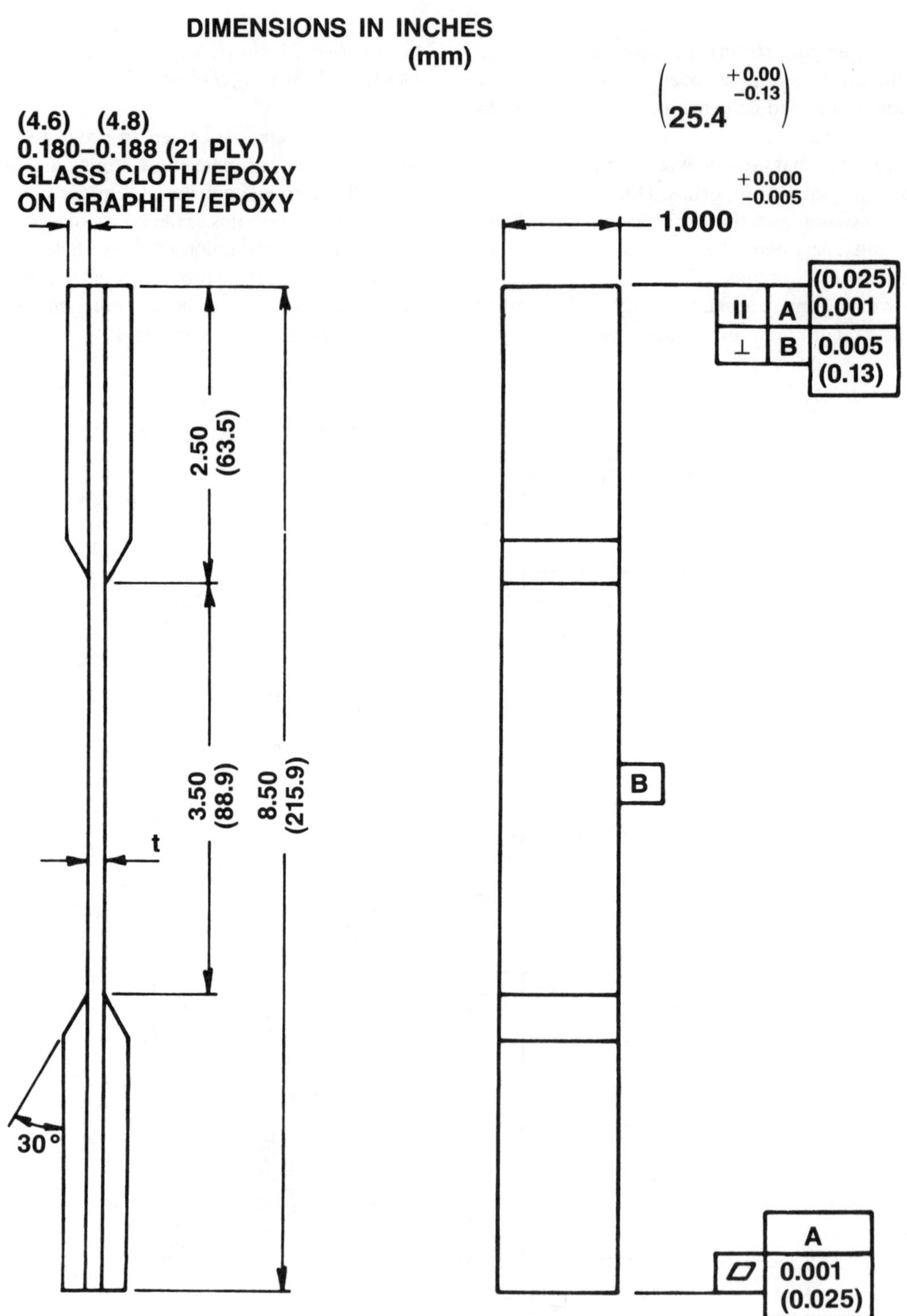

FIGURE 6.1-29b. SWRI fully supported compression specimen.

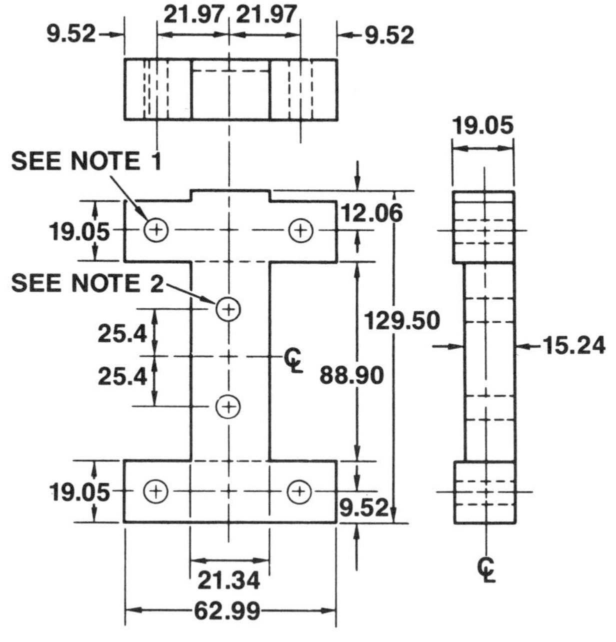

NOTE:

1. TWO REQUIRED: (1) ASSEMBLY DRILLED AND TAPPED 1/4″ -20 4 PLACES, (2) ASSEMBLY DRILLED ONLY
2. HOLE DIAMETER AS REQUIRED FOR EXTENSOMETER AND/OR STRAIN GAGES
3. MATERIAL 1100–1240 MPa STEEL
4. ALL DIMENSIONS IN MILLIMETERS

FIGURE 6.1-30. Lockheed fully supported compression fixture.

FIGURE 6.1-31. Sandwich edgewise compression test specimen.

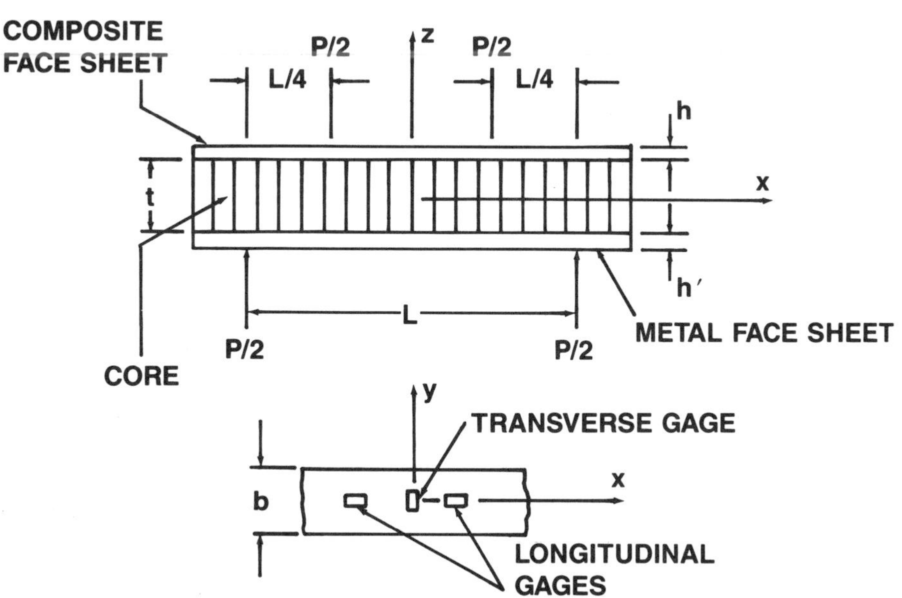

FIGURE 6.1-32. Sandwich beam compression test specimen.

known as *interlaminar shear properties*. To date, test methods have been developed for evaluation of the lamina in-plane properties and for the lamina interlaminar properties in the 1–3 plane. Determination of the interlaminar shear properties in the 2–3 plane is more cumbersome [32]. Shear tests for the composite laminate have generally been restricted to evaluation of properties in the plane of the laminate. No test method has been developed to date that yields intrinsic interlaminar properties for the composite laminate. There are four generally accepted test methods for evaluation of lamina in-plane shear properties: the $[\pm 45]_s$ coupon test, the off-axis coupon, the rail shear test method, and the torsion test. The $[\pm 45]_s$ coupon test method consists of the tensile test of the $[45/-45/45/-45]_s$ laminate of the following geometry:

overall length: 229 mm (9 in)
test section length: 152 mm (6 in)
width: 25 mm (1 in)
top length: 38 mm (1.5 in)

Note that this $\pm 45°$ coupon specimen geometry is identical to that given for the 90° tensile test.

In order to explain how the tensile test of the $[\pm 45]_s$ laminate can be used to evaluate the lamina in-plane shear properties, it will be necessary to review lamination theory (Volume 5). The state of strain for the $[\pm 45]_s$ laminate subjected to a uniaxial tensile stress, $\bar{\sigma}_x = N_x/h$, may be given as follows:

$$\epsilon_x^\circ = H_{11}h\bar{\sigma}_x \qquad K_x = 0$$

$$\epsilon_y^\circ = H_{12}h\bar{\sigma}_x \qquad K_y = 0 \qquad (6.1\text{-}61)$$

$$\gamma_{xy}^\circ = 0 \qquad K_{xy} = 0$$

Therefore, the state of stress within each of the lamina of the laminate may be written in terms of these strains.

$$\sigma_x = \overline{Q}_{11}\epsilon_x + \overline{Q}_{12}\epsilon_y = h\bar{\sigma}_x(H_{11}\overline{Q}_{11} + H_{12}\overline{Q}_{12})$$

$$\sigma_y = \overline{Q}_{12}\epsilon_x + \overline{Q}_{22}\epsilon_y = h\bar{\sigma}_x(H_{11}\overline{Q}_{12} + H_{12}\overline{Q}_{22})$$

$$\tau_{xy} = \overline{Q}_{16}\epsilon_x + \overline{Q}_{26}\epsilon_y = h\bar{\sigma}_x(H_{11}\overline{Q}_{16} + H_{12}\overline{Q}_{26})$$

$$(6.1\text{-}62)$$

Because the shear coupling stiffness terms are equal and opposite for the $+45°$ and $-45°$ layers, note that

$$\overline{Q}_{16}(45) = -\overline{Q}_{16}(-45)$$

$$(6.1\text{-}63)$$

$$\overline{Q}_{26}(45) = -\overline{Q}_{26}(-45)$$

The shear stresses, τ_{xy} in the $+45°$ and $-45°$, are equal and of opposite sign. Further, it can be shown that for $\theta = \pm 45°$ the normal stress σ_y vanishes and σ_x is equal to the applied stress, $\bar{\sigma}_x$:

$$H_{11}Q_{12} + H_{12}Q_{22} = 0$$

$$(6.1\text{-}64)$$

$$H_{11}Q_{11} + H_{12}Q_{12} = 1/h$$

Therefore, the state of stress within each of the layers is biaxial and consists of stress components σ_x and $\pm \tau_{xy}$. If the stresses are now transformed to the lamina coordinate system, it is possible to determine the state of stress in the lamina coordinate system where the shear response is uncoupled from the normal response.

$$\sigma_1 = m^2\sigma_x + 2mn\tau_{xy}$$

$$\sigma_2 = n^2\sigma_x - 2mn\tau_{xy} \qquad (6.1\text{-}65)$$

$$\tau_{12} = -mn\sigma_x - (m^2 - n^2)\tau_{xy}$$

Now, for $\theta = 45°$, $mn = n^2 = m^2 = 0.5$; therefore, Equation (6.1-64) reduces to the following:

$$\sigma_1 = (\sigma_x + \tau_{xy})/2$$

$$\sigma_2 = (\sigma_x - \tau_{xy})/2 \qquad (6.1\text{-}66)$$

$$\tau_{12} = \pm\sigma_x/2 = \pm\bar{\sigma}_x/2$$

The important result shown in Equation (6.1-65) is that, whereas the normal stresses, σ_1 and σ_2, depend on both the applied stress, $\bar{\sigma}_x$, and the induced shear stress, τ_{xy}, the shear stress, τ_{12}, is a simple function of the applied stress, $\bar{\sigma}_x$. This result is important because the induced shear stress, τ_{xy}, is statically indeterminant; that is, for the nonlinear shear response typical of many composite systems, the magnitude of the induced shear stress cannot be determined. Thus, in regions of nonlinear material response, it is not possible to determine the magnitude of either σ_1 or σ_2 as a simple function of the applied stress, $\bar{\sigma}_x$. Note that for any other angle, θ, the in-plane shear stress, τ_{12}, is also indeterminant.

Consider now transformation of the laminate state of strain $(\epsilon_x^{\,\circ},\ \epsilon_y^{\,\circ})$ to the lamina coordinate system:

$$\epsilon_1 = (\epsilon_x^{\,\circ} + \epsilon_y^{\,\circ})/2$$

$$\epsilon_2 = (\epsilon_x^{\,\circ} + \epsilon_y^{\,\circ})/2 \qquad (6.1\text{-}67)$$

$$\gamma_{12} = -(\epsilon_x^{\,\circ} - \epsilon_y^{\,\circ})$$

Therefore, by performing the tensile test of the $[+45/-45/45/-45]_s$ laminate and monitoring the applied stress, $\bar{\sigma}_x$, and the laminate longitudinal and transverse strains, $\epsilon_x^{\,\circ}$ and $\epsilon_y^{\,\circ}$, it is possible to establish the lamina in-plane shear response:

$$\tau_{12} = \bar{\sigma}_x/2$$

$$\qquad (6.1\text{-}68)$$

$$\gamma_{12} = (\epsilon_x^{\,\circ} + \epsilon_y^{\,\circ})$$

Further, the laminate effective modulus, E_x, and Poisson's ratio are defined as follows:

$$E_x = \bar{\sigma}_x/\epsilon_x^{\,\circ}$$

$$\qquad (6.1\text{-}69)$$

$$\nu_{xy} = -\epsilon_y^{\,\circ}/\epsilon_x^{\,\circ}$$

An expression for the in-plane shear modulus, G_{12}, in terms of the $[\pm45]_s$ laminate effective properties may be developed:

$$G_{12} = \frac{E_x}{2(1 + \nu_{xy})} \qquad (6.1\text{-}70)$$

Typical stress-strain results for the $[+45/-45/45/-45]_s$ laminate tensile test of a boron/epoxy material system are shown in Figure 6.1-33. Transformation of these results to lamina in-plane shear response by Equation (6.1-68) is shown in Figure 6.1-34. This test procedure was suggested by Rosen [33], and details are presented in ASTM Standard D3518-76.

It should be noted that, although the $[\pm45]_s$ laminate tensile test can be employed to establish shear stress-strain response well into the region of nonlinear material response [34], caution must be exercised in interpreting the ultimate stress and strain results. As

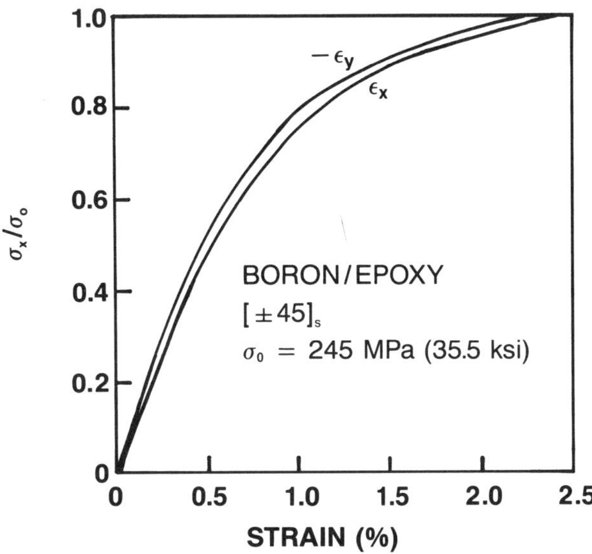

FIGURE 6.1-33. Stress-strain for $[\pm45]_s$ boron/epoxy laminate.

shown previously, this caution is due to the fact that the lamina state of stress is not pure shear but rather biaxial in nature. Hence, it should be expected that the presence of the normal stress components would have a deleterious effect upon shear strength. In particular, the transverse normal stress, σ_2, can be expected to significantly alter the apparent shear strength. Data that

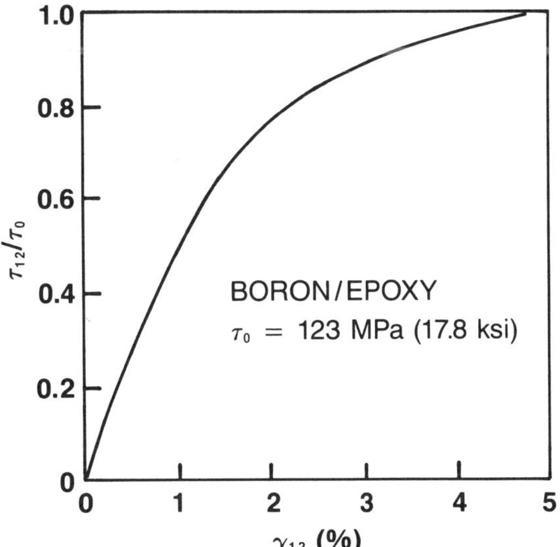

FIGURE 6.1-34. Shear stress-strain curve for unidirectional boron/epoxy composite as derived from Figure 6.1-33.

Table 6.1-1. Graphite/epoxy $[\pm 45_{16}]_s$ laminate test results.

Tension MPa (ksi)	Compression MPa (ksi)
172 (24.9)	203 (29.4)
177 (25.6)	206 (29.9)
163 (23.6)	193 (27.9)
155 (22.5)*	204 (29.5)
148 (21.5)*	202 (29.3)
144 (20.9)*	203 (29.4)
152 (22.0)*	
155 (22.5)*	
148 (21.5)*	
Average:	Average:
157 (22.8)	201 (29.2)

*8-ply laminate, $[\pm 45_2]_s$.

tend to confirm the expected influence of transverse normal stress upon shear strength were developed by comparing the ultimate strengths of the $[\pm 45]_s$ laminate in tension and compression. When the laminate is loaded in tension, the normal stress components are tensile, whereas a compression loading leads to compressive normal stresses. Because the presence of a tensile transverse normal stress is expected to *reduce* the apparent in-plane shear strength whereas a compressive normal stress may result in an *increase* in strength, a compressive strength larger than the tensile strengths of the $[\pm 45]_s$ laminate would support the argument of the deleterious influence of the tensile transverse stress upon ultimate shear strength. Such data are presented in Table 6.1-1 for a graphite/epoxy composite where it can be seen that the average tensile strength for the $[\pm 45]_s$ laminate is 157 MPa (22.8 ksi), while the average compressive strength is 201 MPa (29.2 ksi).

The second method for determination of lamina shear response is the off-axis tensile test, which was discussed in subsection 6.1.4. For shear response relative to the fiber direction under the uniaxial load, σ_x, Equations (6.1-51) and (6.1-55) yield

$$\tau_{12} = -mn\sigma_x \qquad (6.1\text{-}71)$$

$$\gamma_{12} = -(m^2 + 2mn - n^2)\epsilon_x - (m^2 - 2mn - n^2)\epsilon_y$$

$$+ 2(m^2 - n^2)\epsilon_{45} \qquad (6.1\text{-}72)$$

As previously discussed, Equation (6.1-72) is based on the use of a rectangular three-element rosette gage. If

Equations (6.1-71) and (6.1-72) are specialized for $\theta = 45°$, then $m^2 = n^2 = mn = 0.5$ and

$$\tau_{12} = \frac{-\sigma_x}{2} \qquad (6.1\text{-}73)$$

$$\gamma_{12} = \epsilon_y - \epsilon_x \qquad (6.1\text{-}74)$$

Therefore,

$$\gamma_{12} = -(1 + \nu_{xy})\frac{\sigma_x}{E_x} \qquad (6.1\text{-}75)$$

Combining Equations (6.1-73) and (6.1-75) allows determination of the lamina in-plane shear modulus, G_{12}, in terms of the 45° off-axis laminate properties:

$$G_{12} = \frac{\tau_{12}}{\gamma_{12}} = \frac{E_x}{2(1 + \nu_{xy})} \qquad (6.1\text{-}76)$$

It should be noted that Equation (6.1-76) is identical to Equation (6.1-70) developed for the $[\pm 45]_s$ tensile test. The response of the 45° off-axis and $[\pm 45]_s$ laminate test specimens is, however, quite different. For example, the effective properties of the two test specimens for a given boron/epoxy material are given as follows:

$$E_x(\pm 45) = 21.0 \text{ GPa (3.04 Msi)}$$

$$\nu_{xy}(\pm 45) = 0.69$$

$$G_{12} = \frac{(20.98)}{2(1 + 0.69)} = 6.21 \text{ GPa (0.90 Msi)}$$

$$E_x(45) = 15.9 \text{ GPa (2.30 Msi)}$$

$$\nu_{xy}(45) = 0.28$$

$$G_{12} = \frac{(15.9)}{2(1 + 0.28)} = 6.21 \text{ GPa (0.90 Msi)}$$

Note that both of the effective elastic properties for the $[\pm 45]_s$ laminate test are greater than the corresponding properties of the 45° off-axis laminate, yet the same value of G_{12} is determined by each of the test methods. As previously discussed, Equation (6.1-49) can be utilized in designing the dimensions of the off-axis tensile specimen to assure that the shear coupling effects are minimized (see also reference [28]).

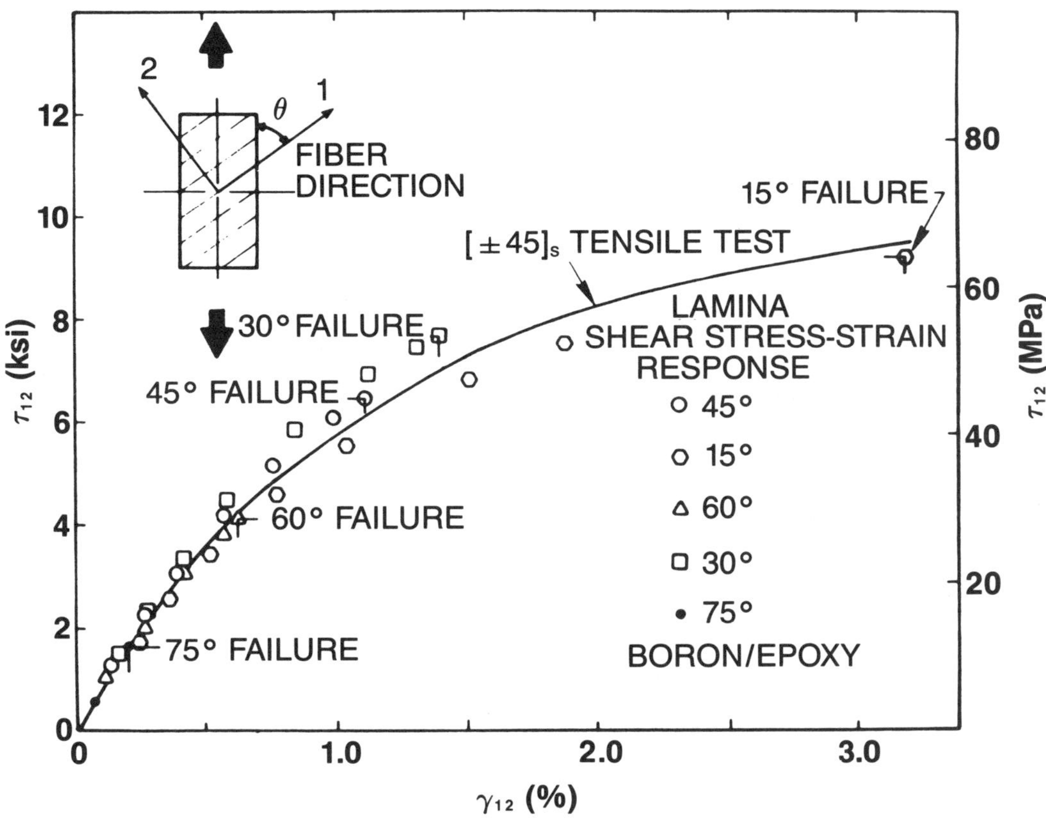

FIGURE 6.1-35. Off-axis test for shear properties.

The in-plane shear stress-strain response, as determined in the off-axis test, is consistent for any angle, θ, as shown in Figure 6.1-35 for boron/epoxy. As noted in this figure, however, the shear strength is a function of the fiber orientation of the off-axis specimen. Although the analysis presented earlier indicates that the 45° off-axis specimen is attractive because the relation between the lamina shear response and the 45° off-axis lamina response is quite simple, the results presented in Figure 6.1-35 show that ultimate shear strength predicted by the 45° specimen may be in error by as much as 30–40 percent. Hence, this error led Chamis and Sinclair [35] to recommend the 10° off-axis test for determining lamina shear properties. The 10° angle was chosen to minimize the effects of longitudinal and transverse tension components, σ_1 and σ_2, on the shear response. Results for the 15° off-axis test shown in Figure 6.1-35 indicate that the ultimate shear strengths determined by the 15° test and the $[\pm45]_s$ tensile test compare quite

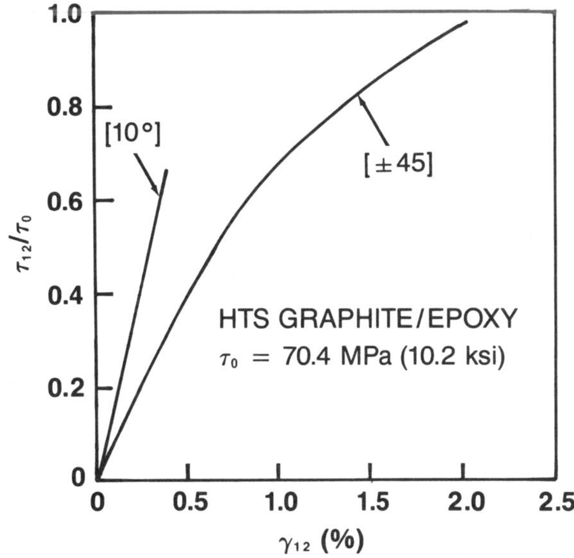

FIGURE 6.1-36. 10° off-axis tensile test results.

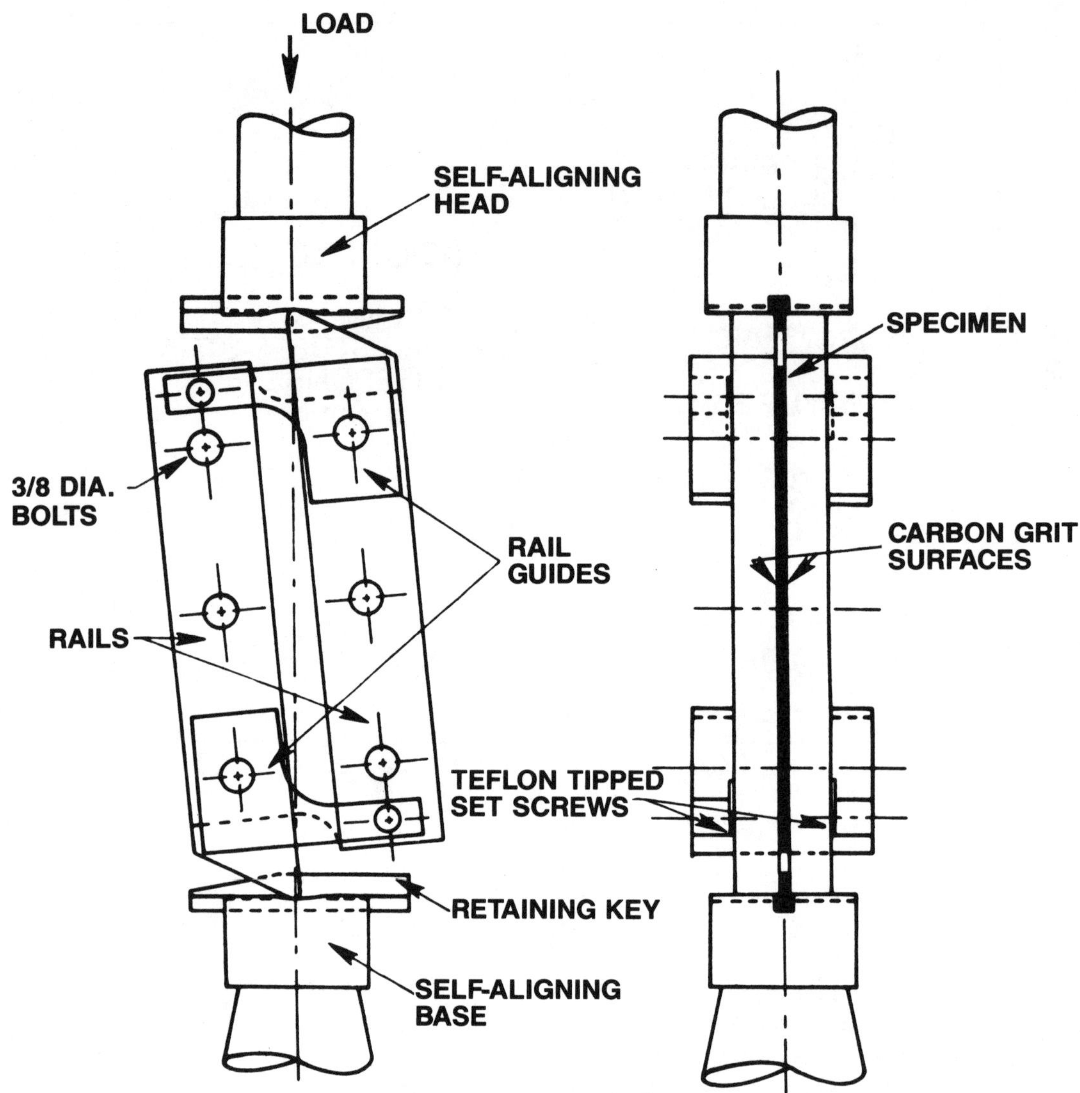

FIGURE 6.1-37. Two-rail shear test.

favorably. A comparison between the 10° off-axis method and the [±45]ₛ laminate method for determining shear stress-strain behavior is shown in Figure 6.1-36 for graphite/epoxy. Note that the 10° off-axis test yields a higher initial modulus, while the [±45]ₛ laminate test yields a higher strength. The [±45]ₛ laminate test yields considerably more of the shear stress-strain curve than the 10° off-axis test.

The third method for determination of lamina in-plane shear response is the rail shear test method. Two test specimens and fixture configurations are generally accepted: the two-rail fixtures shown in Figure 6.1-37 and the three-rail fixtures shown in Figure 6.1-38. The test fixture consists primarily of rigid steel rails that are bolted or bonded to a rectangular laminate specimen. For determination of lamina shear properties, the laminate consists of eight to twelve plies whose fiber orientation is parallel and/or perpendicular to the longitudinal axis of the rails. It should be noted, however, for the case where the fibers are perpendicular to the rails, the clamping of the fibers may induce a higher shear strength compared to the case where the fibers are parallel to the rails. A [0/90]ₛ laminate may also be utilized for unidirectional characterization, since the shear response should be identical for 0° and 90° orientations (i.e., $A_{66} = G_{12}h$ for [0/90]$_s$ laminates). Laminates having orthotropic in-plane properties ($A_{16} = A_{26} = 0$) relative to an *x-y* axis system parallel and perpendicular to the rails, respectively, may also be utilized in this method.

The test specimen geometries for the two-rail and three-rail configurations are shown in Figures 6.1-39 and 6.1-40, respectively. The shear stress calculation for the two configurations may be expressed as a function of the applied load, P, the laminate thickness, h, and the width between the rails, b:

$$\bar{\tau}_{xy} = \frac{N_{xy}}{h} = \frac{P}{bh} \qquad \text{(two-rail)}$$

$$\text{(6.1-77)}$$

$$\bar{\tau}_{xy} = \frac{N_{xy}}{h} = \frac{P}{2bh} \qquad \text{(three-rail)}$$

Shear strain can be measured with a rectangular rosette

gage located at the geometric center of the specimen. For the case of pure shear,

$$\epsilon_x = \epsilon_y = 0 \qquad \text{(6.1-78)}$$

Equation (6.1-52) yields,

$$\gamma_{xy} = 2\epsilon_{45} \qquad \text{(6.1-79)}$$

Thus, in theory, a single element gage at 45° is sufficient to determine shear strain. The three-element gage, however, serves as a means of assuring that the strain field in the center of the specimen is pure shear (i.e., the longitudinal and transverse gages should show essentially zero strain). It should be noted that the free edges at the top and bottom of the rail shear specimen experience large normal stresses concentrated in the corners [1]. If the length-to-width ratio of the gage section is sufficiently large (8:1), a state of uniform shear will exist in the center of the gage section. These results

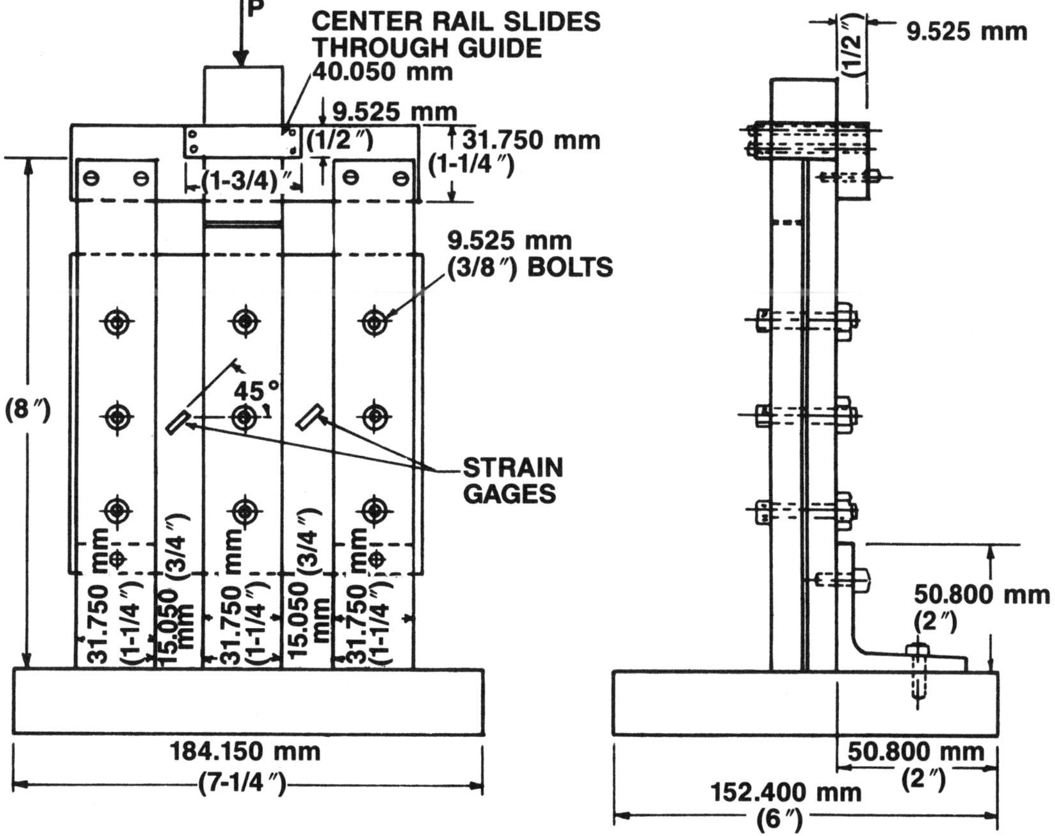

FIGURE 6.1-38. Three-rail shear test.

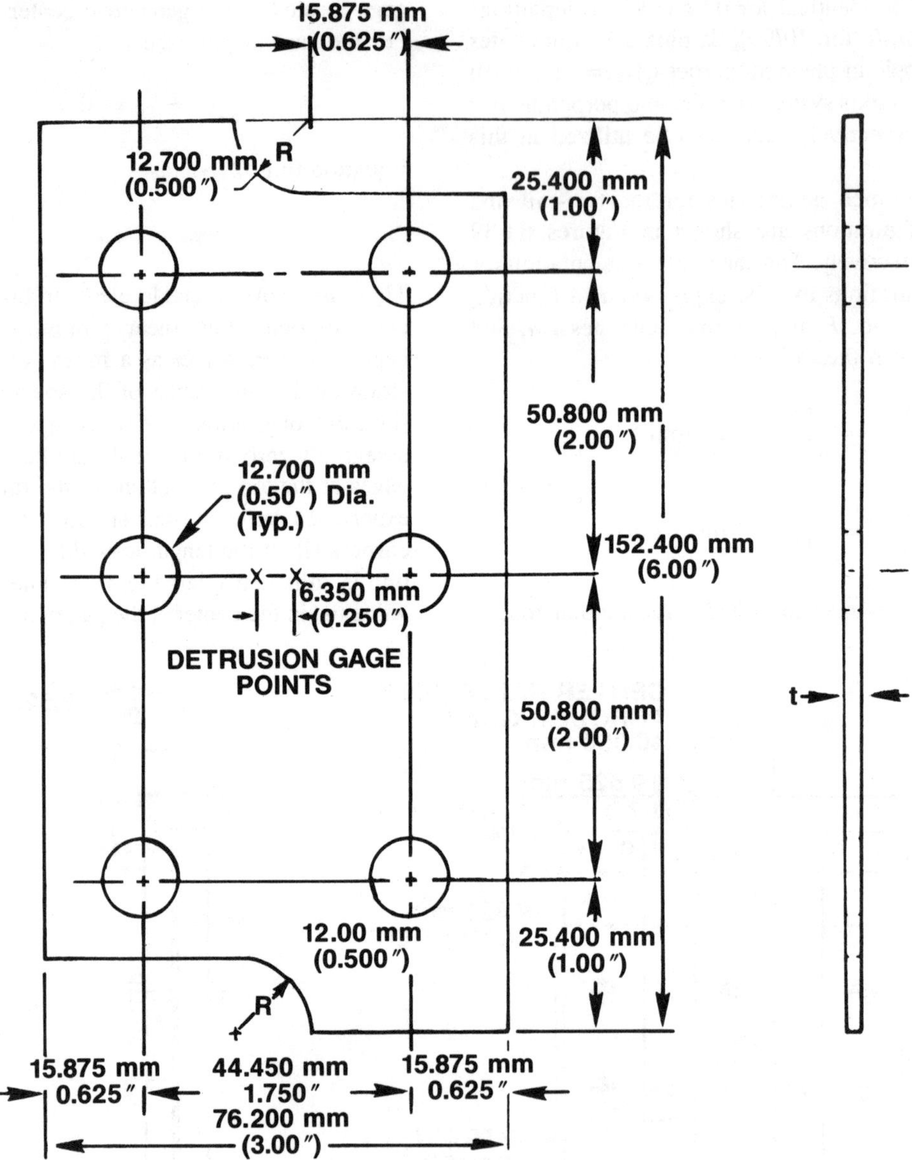

FIGURE 6.1-39. Two-rail shear test specimen.

are obtained on the basis that the edges are perfectly clamped. Thus, a means of checking the strain field in the center of the gage section is desirable. The perfectly clamped boundary assumption requires that the bolts in the rail shear specimens apply even clamping pressure to the edges. The laminate shear modulus, $G_{xy} = A_{66}/h$, can be determined from the slope of the initial straight line portion of the shear stress-strain curve.

Another method for determining in-plane shear properties of a laminate is the torsion method, which can be utilized in conjunction with either a solid rod (unidirectional specimen only) or a tubular specimen.

In the solid rod method, a unidirectional specimen is fabricated by either laying up the material in a cylindrical mold or machining the specimen from a solid rectangular bar. This test is not widely used, however, because other methods are better. In particular, a solid rod under torsion yields a shear stress distribution which is linear with respect to the radial distance from the center of the rod. Thus, the calculation for stress is accurate only for linear portions of the load-deformation curve. With the large nonlinear response observed for unidirectional shear stress-strain curves, the usefulness of the test is highly reduced. In addition, the solid rod specimen is difficult to fabricate. Other approaches, such as the $[\pm 45]_s$ tensile test, do not have these drawbacks.

Torsion of a thin-walled tube provides a means of directly applying pure shear to a fiber reinforced composite specimen. The major drawback to this method, however, is the cost and difficulty associated with fabricating quality tubular specimens. In addition, the method requires specialized equipment and gripping systems. Problems associated with end fixtures are discussed in greater detail in conjunction with a later segment on combined loading of tubular specimens. For the case of pure torsion, the end attachment problem is much less severe than for the case of combined loading. A simple and effective grip design involves metal grips bonded to the tube ends with an epoxy potting material. The grips are bonded on both the outside and inside of the tube to increase the area of load introduction. There are a number of design configurations that these grips can take. One of the most efficient from both a load introduction and specimen alignment standpoint is the grip design of Hahn and Erikson [36], as shown in

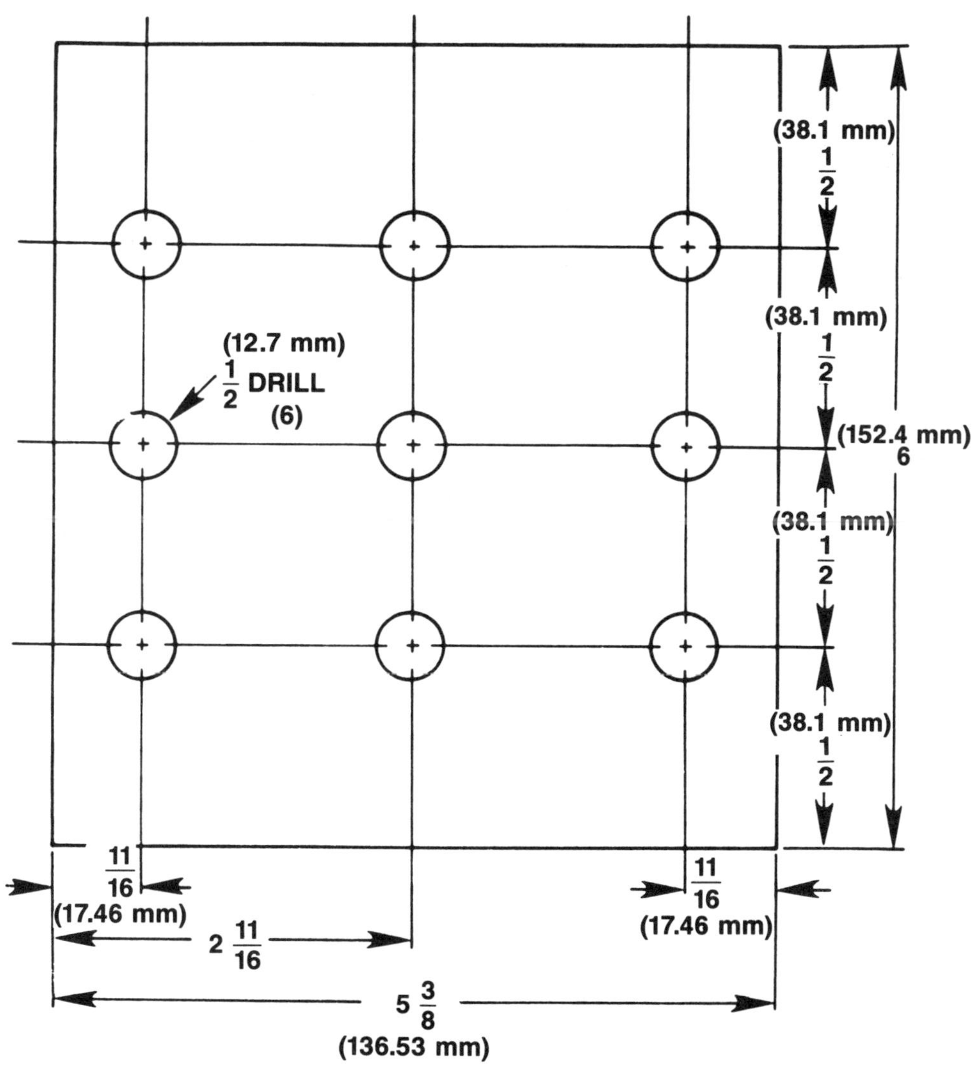

FIGURE 6.1-40. Three-rail shear test specimen.

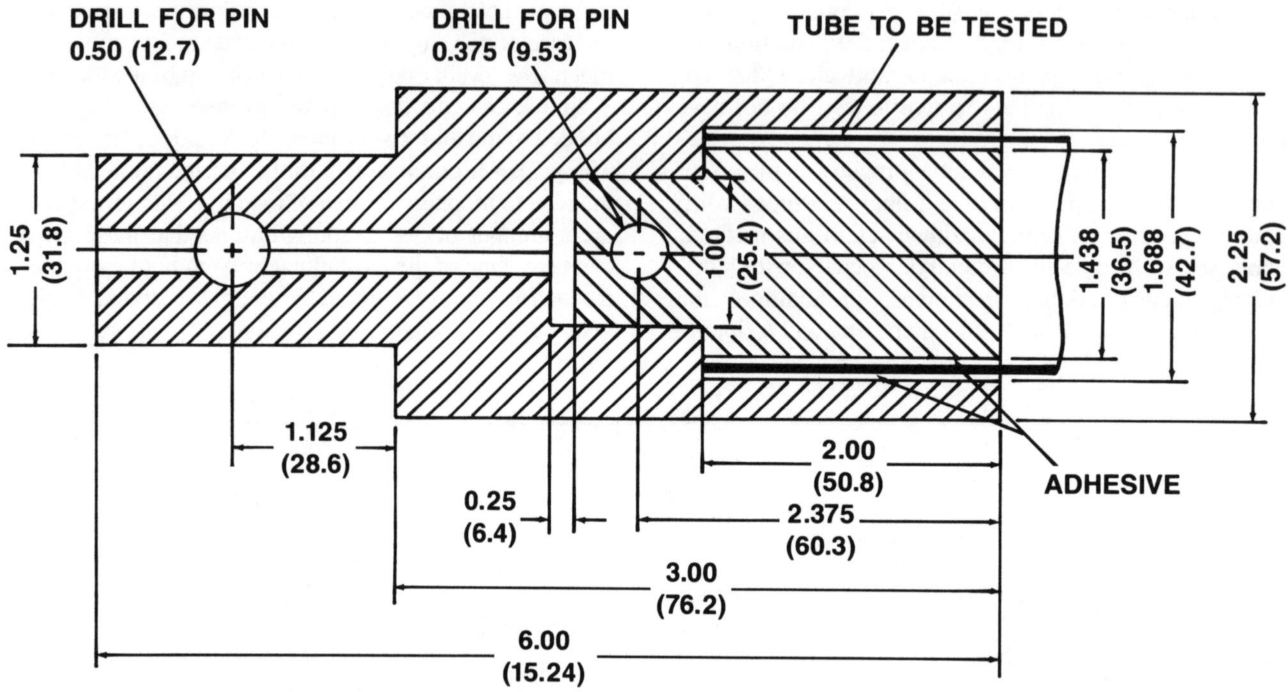

FIGURE 6.1-41. Tubular specimen grips.

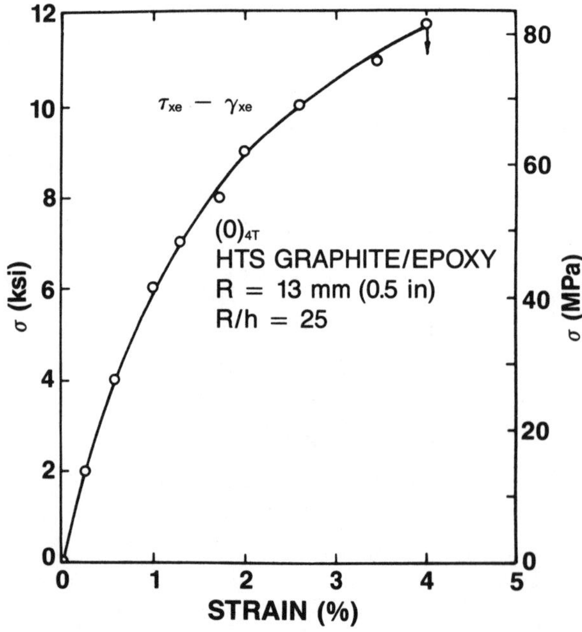

FIGURE 6.1-42. 0° tubular specimen stress-strain results.

Figure 6.1-41. This grip consists of concentric cylinders with the tubular specimen bonded between the cylinders. Two pins through the outer cylinder are used in conjunction with the fixture. One pin connects the inner cylinder with the outer cylinder, and the second connects the entire fixture and specimen to the loading machine. This type of fixture has the advantage of not requiring reinforced ends.

For the case of orthotropic laminates ($A_{16} = A_{26} = 0$) of symmetric construction ($B_{ij} = 0$) subjected to pure torsion,

$$\bar{\tau}_{xy} = \frac{N_{xy}}{h} = \frac{T}{2\pi R^2 h} \qquad (6.1\text{-}80)$$

where T is the applied torque and R is the mean radius of the cylinder. The shear strain is related to the angle-of-twist per unit length, ϕ, by

$$\gamma^{\circ}_{xy} = R\phi \qquad (6.1\text{-}81)$$

Thus, a shear stress-strain curve can be obtained from a plot of torque versus angle-of-twist. The angle-of-twist, however, contains effects of machine compliance and is not an accurate measure of shear strain. Much more accurate results can be obtained by using a strain gage. As in the case of rail shear, a rectangular rosette gage provides a direct measurement of shear strain; i.e., Equation (6.1-79) can be used to determine the shear strain in the tube. Again, the longitudinal and transverse gages should register zero for pure shear. It should also be noted that a strain gradient exists in a composite tubular specimen. This gradient can be approximated as a linear function of the z coordinate located in the center of the tube wall and directed normal to the outer surface of the cylinder [37]:

$$\gamma_{xy} = \left(\frac{1 + z}{R} \right) \gamma_{xy}^{\circ} \qquad (6.1\text{-}82)$$

This gradient can be checked experimentally by using gages both inside and outside of the tube. If the shear strain is measured on the outer surface, then Equation (6.1-82) yields

$$\gamma_{xy}^{\circ} = \frac{\gamma_{xy}}{(1 + 2h/R)} \qquad (6.1\text{-}83)$$

In order to minimize the effect of the strain gradient on the ply stresses, $R/h \geq 10$. In addition, to insure sufficient gage section in the tube, $L/R \geq 8$, where L denotes the length of the tube between end fixtures.

The composite shear modulus, $G_{xy} = A_{66}/h$, can be obtained by measuring the slope of the initial straight line portion of the $\bar{\tau}_{xy} - \gamma_{xy}^{\circ}$ stress-strain curve.

Shear stress-strain curves are shown in Figure 6.1-42 for a $[0]_{4T}$ unidirectional tube and in Figure 6.1-43 for a $[0/90]_s$ laminated tube fabricated from HTS graphite/epoxy resin. As predicted from laminated plate theory, both of the shear stress-strain curves are almost identical.

Unlike the shear test methods discussed previously, the short beam shear method is utilized to estimate interlaminar shear strength only [1]. This method does not provide any measurement of shear modulus. Because of its simplicity, the short beam shear test is often used as a quality control tool.

The short beam shear test involves a three-point flexure specimen with the span-to-depth ratio, L/h, chosen to produce interlaminar shear failure. This method is subject to the same restrictions with regard to material applicability as discussed in subsection 6.1.4 for the general flexure test. An additional complexity is presented by the short beam shear method when used in conjunction with laminated materials. In particular, the interlaminar shear stress will be parabolic within each layer, but a discontinuity in slope will occur at the ply interfaces. As a result, the maximum shear stress will not necessarily occur at the center of the beam, as shown for the $[0/90]_s$ graphite/epoxy laminate in Figure 6.1-44.

The short beam method is commonly applied only to polymeric and composite materials that can be treated as homogeneous. For three-point bending, the interlaminar shear strength is determined from the relationship

$$\tau_m = \frac{3P}{4bh} \qquad (6.1\text{-}84)$$

where τ_m is the maximum value of τ_{xz}. Using Equation (6.1-29), it can easily be seen that a shear failure should be assured if

$$L/h < s_x/s_{xz} \qquad (6.1\text{-}85)$$

where s_x and s_{xz} are the tensile strength (or compressive

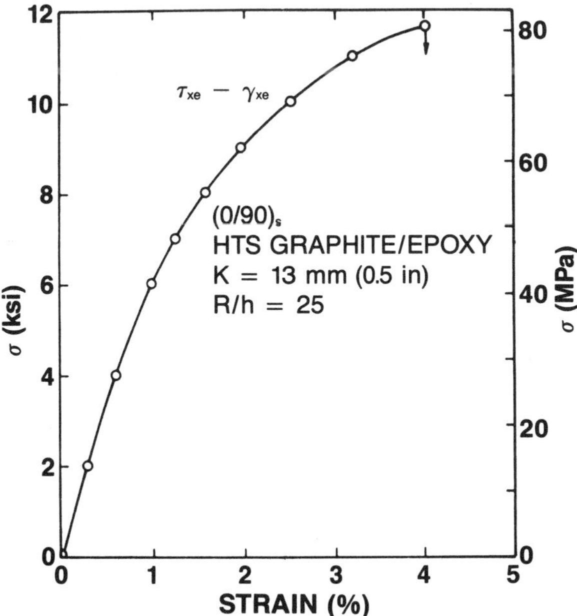

FIGURE 6.1-43. $[0/90]_s$ tubular specimen stress-strain results.

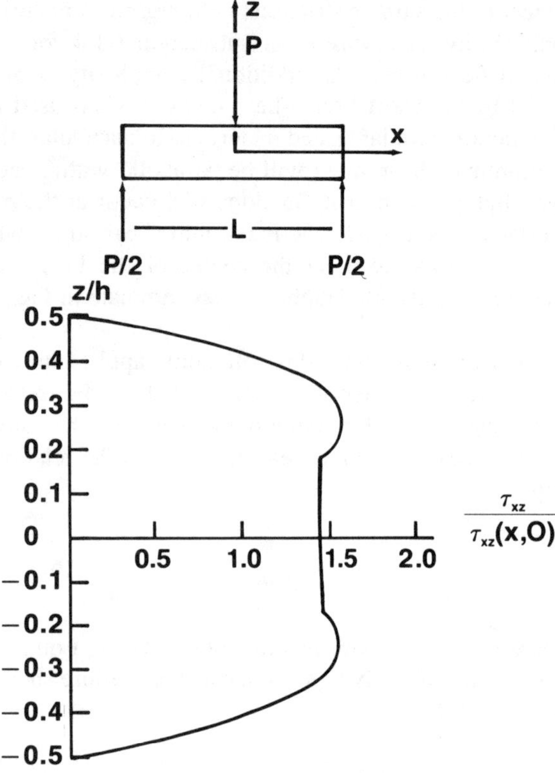

FIGURE 6.1-44. Shear beam stress distribution $[0/90]_s$.

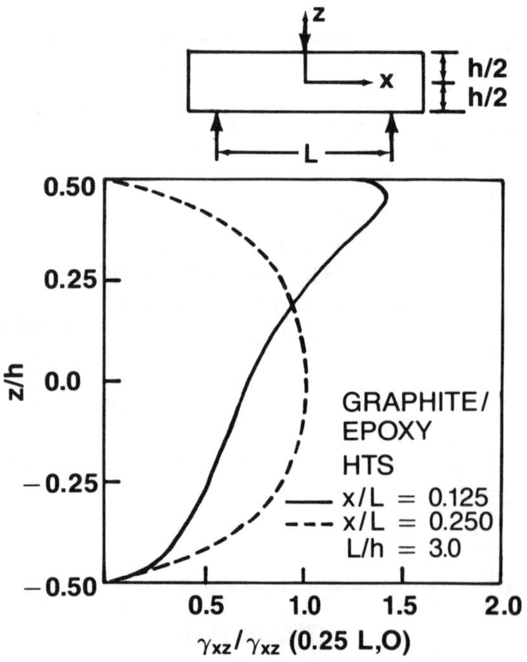

FIGURE 6.1-45. Shear strain distribution at $X/L = 0.083$ and 0.250.

strength, whichever is the lesser) and interlaminar shear strength, respectively.

Precise dimensions recommended for the short beam shear test can be found in ASTM D2344-76. It should be noted that recommended values of $4h$ are considerably less than the maximum values as determined from Equation (6.1-85). For example, a recommended value of $L/h = 4$ for graphite/epoxy, $s_x > 1,400$ MPa (200 ksi), is much less than required by the inequality in Equation (6.1-85). Practical experience has shown, however, that interlaminar shear failures are difficult to attain at higher span-to-depth ratios. A maximum thickness of 6.4 mm (0.25 in) is required by the ASTM standard. There is no minimum thickness required, and, as a result, short beam shear specimens as thin as 1.5 mm (0.06 in) have been tested. This means that, at $L/h = 4$, a load nose of 6.4 mm (0.25 in) diameter will cover the entire span length creating a "punchout" test rather than a flexure test. Test specimens should be as thick as possible to avoid this problem.

In order to avoid these difficulties, including measurement of interlaminar shear stiffness, a very large interlaminar shear beam was investigated by Pipes, Reed, and Ashton [38]. These beams were 13 mm (0.5 in) in width, 25 mm (1 in) thick, and 102 mm (4 in) long. Although this approach alleviates the problem associated with excessively small values of L resulting from low span-to-depth ratios, the problem of large shear stresses in excess of those predicted by classical beam theory is present. A finite element analysis performed by Berg, Tirosh, and Israeli [39] showed such large shear stresses in the vicinity of the applied center load. Similar results were obtained by Pipes, Reed, and Ashton [38], as shown in Figure 6.1-45. Note, at $x/L = 0.25$, classical beam shear strains are recovered. These large shear strains and locations were experimentally verified using photoelastic coatings.

The difficulties associated with the short beam shear test have not influenced its popularity. Simplicity of testing and the inexpensive nature of the specimen make the method an ideal tool for quality control or materials screening. Despite its wide use, the value of the short beam shear test remains highly questionable.

Biaxial Loading

As discussed previously, the off-axis tensile test can be used as a means of measuring biaxial stress-strain behavior relative to the fiber direction [40]. This method could be extended to include orthotropic laminates in an off-axis orientation. This method, however, has a number of drawbacks, with the most significant being the inability to independently control the biaxial stress components. Once a uniaxial tensile test method is discarded, alternate approaches require special equipment and fixtures.

An ideal biaxial test method must meet the following requirements:

a) A significant volume of material must be under a state of uniform stress (σ_x, σ_y, τ_{xy}).
b) Primary failure must occur in the test section.
c) The state of stress must be known without secondary measurements or analysis.
d) It must be possible to vary the stress components independently.

A thin-walled tubular specimen under combined axial load, torsion, and internal pressure meets these requirements. As mentioned, this method is hindered by difficulties associated with fabricating quality tubular specimens and by difficulties associated with load introduction.

A common method for fabricating tubular specimens utilizes a steel cavity tool, a hollow, perforated mandrel, and an elastomer bladder. The tubular laminate is internally pressurized so that it conforms to the shape of the cylindrical cavity. Complete details on such a procedure can be found in reference [41]. Fiber volume content and uniformity of wall thickness are two of the most difficult parameters to control.

Axial loads and internal pressure present a particular problem in terms of end attachments. In particular, when the transverse and radial displacements of a tubular specimen are restricted by end attachments, extraneous bending stresses are induced in tubular regions near the end attachments. Such stress concentrations have been confirmed analytically and experimentally for tubular specimens under axial load. Laminates that have very large effective Poisson's ratios, such as $[\pm 45]_s$ laminates, are a particular problem. These stress and strain concentrations may induce premature end failures, and thus render the ultimate strength results meaningless. Bonded grips, as discussed previously, allow some radial dimensional change; as a result, they can be utilized for many laminates and load combinations. Self-compensating end grips have also been developed [41]. Such designs utilize combinations of internal and external pressure in the end grip area to provide radial displacements that match gage section displacements. Strain gages can be used to sense any

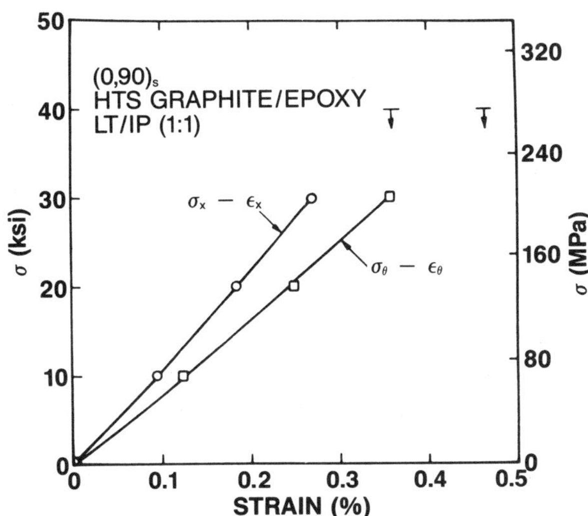

FIGURE 6.1-46. Biaxial tension stress-strain response $[0/90]_s$.

strain gradient and automatically trigger the self-compensating device. Grips of this nature are very expensive and require peripheral electronic equipment.

Consider a tube with a Cartesian coordinate system located in the middle surface of the tube with x, y, and z measured along the longitudinal, circumferential, and radial direction, respectively. For any combination of axial load, internal pressure, and torque, the average stresses in the tube are given by

$$\bar{\sigma}_x = \frac{N_x}{h}, \ \bar{\sigma}_y = \frac{N_y}{h} = \frac{R}{h}p,$$

$$\bar{\tau}_{xy} = \frac{T}{2\pi h R^2} \tag{6.1-86}$$

where h is the thickness of the tube wall, R is the radius of the middle surface of the tube, p is the internal pressure, and T is the torque. The constitutive equations of the cylinder when the moments vanish are

$$\epsilon_x^\circ = H_{11}h\bar{\sigma}_x + H_{12}Rp + H_{16}\frac{T}{2\pi R^2} \tag{6.1-87}$$

$$\epsilon_y^\circ = H_{12}h\bar{\sigma}_x + H_{22}Rp + H_{26}\frac{T}{2\pi R^2} \tag{6.1-88}$$

$$\gamma_{xy}^\circ = H_{16}h\bar{\sigma}_x + H_{26}Rp + H_{66}\frac{T}{2\pi R^2} \tag{6.1-89}$$

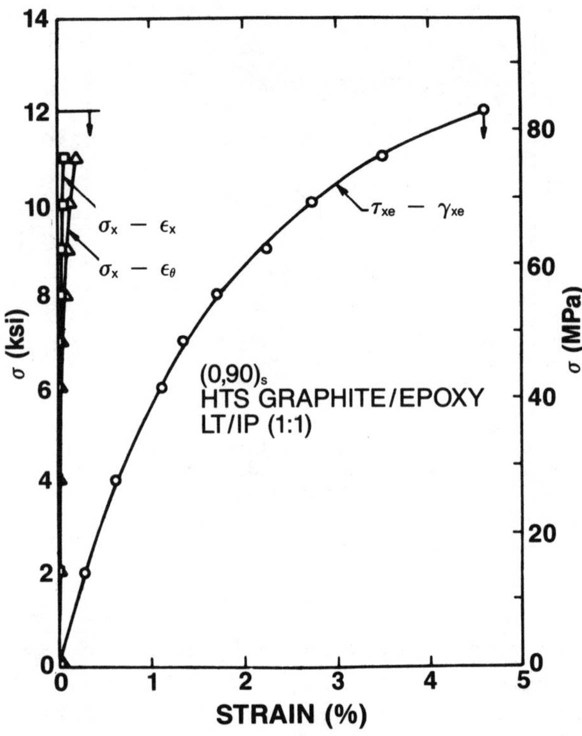

FIGURE 6.1-47. Biaxial stress-strain response $[0/90]_s$.

The effective elastic constants H_{ij} can be determined from A_{ij} by application of lamination theory (Volume 5).

For anisotropic tubes ($A_{16} = A_{26} \neq 0$), Equation (6.1-89) reveals that a shear strain will be induced by axial tension. In such cases, one end of the tube must be allowed to rotate or shear stress is induced. Under certain circumstances shear buckling can result [42]. A rectangular, three-element rosette gage can be used to measure outer surface strains that are related to mid-plane strains in the following manner [43]:

$$\epsilon_x^\circ = \epsilon_x \qquad (6.1\text{-}90)$$

$$\epsilon_y^\circ = \frac{\epsilon_y}{(1 + h/2R)} \qquad (6.1\text{-}91)$$

$$\gamma_{xy}^\circ = \frac{\gamma_{xy}}{(1 + h/2R)} \qquad (6.1\text{-}92)$$

Dimensions of the tube should be chosen to minimize strain gradient effects and to assure uniform stresses in the gage section. The former can be accomplished by

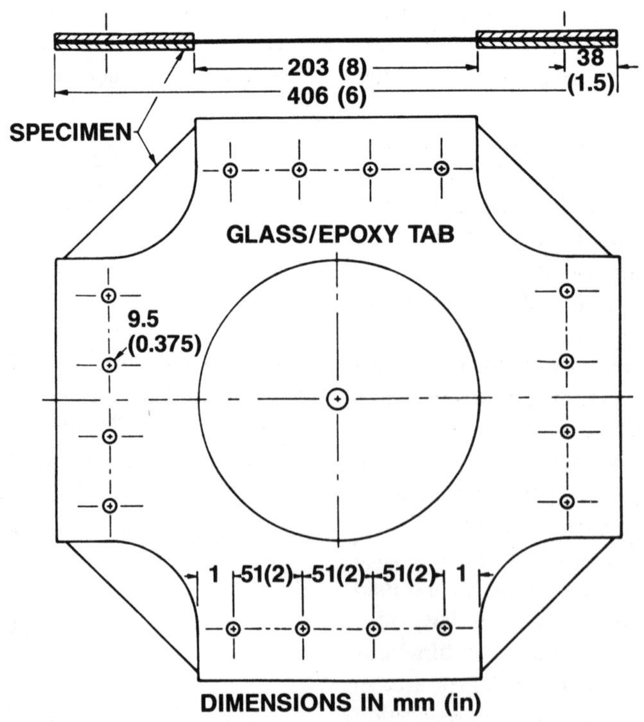

FIGURE 6.1-48. IITRI flat laminate biaxial specimen.

FIGURE 6.1-49. IITRI biaxial test.

making $R/h \geq 10$, and the latter can be assured by utilizing the relationship

$$L = 4R + \bar{L} \qquad (6.1\text{-}93)$$

where L denotes total tube length and $\bar{L}$ denotes the desired tube gage length.

Stress-strain curves are shown in Figures 6.1-46 and 6.1-47 for combined loading of a $[0/90]_s$ laminate tube fabricated of HTS graphite/epoxy. Longitudinal tension (LT) and internal pressure (IP), one-to-one ratio, are shown in Figure 6.1-46. Combined longitudinal tension and torsion (T), one-to-one ratio, are shown in Figure 6.1-47. Note that in Figure 6.1-47, the longitudinal tension has little effect on the shear stress-strain response as seen by comparing Figure 6.1-47 with the pure shear results shown in Figure 6.1-43.

While the tubular specimen is ideal for measuring combined load response for unnotched composites, it is unacceptable for determining the biaxial response of notched composites. Evaluation of composites containing stress concentrations, such as circular holes, slits, and cutouts, requires the use of flat plate specimens [1,2]. For combined loading, such an evaluation requires rather large specimens and specialized test fixtures.

A flat plate biaxial test has been developed by Daniel [44,45]. This method utilizes 8-ply specimens 40 cm $\times$ 40 cm (16 in $\times$ 16 in). Corners of approximately 5 cm (2 in) sides are cut off from the composite laminates. They are then tabbed with 5-ply crossply glass/epoxy tabs. These tabs contain a circular cutout in the center of 20.3 cm (8 in) diameter. A sketch of this specimen is shown in Figure 6.1-48. Deformations and strains are obtained from strain gages. For Daniel's studies [44,45], gages were mounted on the hole boundary, near it, and in the far field along the horizontal (x-) and vertical (y-) axes of material symmetry. In most cases birefringent coatings 0.5 mm (0.02 in) and 1 mm (0.04 in) thick were used on one side of the specimen.

Four 0.95 cm (0.037 in) diameter holes are provided on each side of the tabbed specimen for bolting individual pairs of metal grips approximately 5 cm (2 in) wide and 10 cm (4 in) long. Loading is introduced by means of four whiffle-tree grip linkages designed to ensure that four equal loads are applied to each side of the specimen. A photograph of a biaxial specimen with the loading grip linkages is shown in Figure 6.1-49. Note

that the tab geometry and slits are an early design, which was found to be unacceptable due to stress concentrations. The grip linkage concept, however, was retained in later designs.

Load is applied by means of two pairs of hydraulic jacks attached to the four sides of a reaction frame. The load is transmitted from the hydraulic cylinders to the grip linkages through the bore of these cylinders. The rods are instrumented with strain gages and calibrated in a testing machine to establish the exact relationship between loads and strain gage signals. These strain gage readings are used subsequently both for recording the exact loads applied to the specimen and as feedback signals for controlling the pressures by means of the servohydraulic systems used.

The relationship between the input load and the effective stress for the specimen geometry used and stress biaxiality desired must be established by preliminary calibration tests. Unnotched specimens tabbed as previously described and instrumented with strain gages at the center and far-field locations can be used to perform the calibration. For an orthotropic laminate ($A_{15} = A_{26} = 0$), the constitutive relations yield

$$\epsilon_x = \frac{N_y}{hE_x}(B - \nu_{xy}) \qquad (6.1\text{-}94)$$

$$\epsilon_y = \frac{N_y}{hE_x}\left(\frac{E_x}{E_y} - \nu_{xy}B\right) \qquad (6.1\text{-}95)$$

where B is the biaxiality ratio, N_x/N_y. Combining Equations (6.1-94) and (6.1-95) yields

$$\frac{\epsilon_x}{\epsilon_y} = \frac{(B - \nu_{xy})}{(E_x/E_y - \nu_{xy}B)} \qquad (6.1\text{-}96)$$

Substituting measured values of the elastic constants E_x, E_y, and ν_{xy} along with the desired value of B into Equation (6.1-96) yields ϵ_x/ϵ_y. The input load ratio must be adjusted until the desired strain ratio in Equation (6.1-96) is obtained at the center of the plate.

For laminates containing a stress concentration in the center, the flat plate biaxial test method described appears to yield excellent results. Difficulties are encountered with unnotched specimens, however, as stress concentrations associated with load introduction may lead to premature failure.

6.1.5 References

1. WHITNEY, J. M., I. M. Daniel, and R. B. Pipes. *Experimental Mechanics of Fiber Reinforced Composite Materials, rev. ed.*, Prentice-Hall (1984).

2. CARLSSON, L. A. and R. B. Pipes. *Experimental Characterization of Advanced Composite Materials.* Prentice-Hall (1987).

3. HAHN, H. T. and R. Y. Kim. "Swelling of Composite Laminates," *Advanced Composite Materials—Environmental Effects*, ASTM STP 658, Philadelphia, pp. 98–120 (1978).

4. THOMPKINS, S. S., D. E. Bowles, and W. R. Kennedy. "Laser-Interferometric Dilatometer for Thermal Expansion Measurements of Composites," *Exp. Mechanics*, 26:1–6 (1986).

5. HAHN, H. T. "Residual Stresses in Polymer Matrix Composite Laminates," *J. Composite Materials*, 10(4):266–278 (October 1976).

6. BUECHE, F. *Physical Properties of Polymers.* Interscience Publishers, New York (1962).

7. YANG, P., L. A. Carlsson, and S. S. Sternstein. "Dynamic-Mechanical Response of Graphite/Epoxy Composite Laminates and Neat Resin," *Polymer Composites*, 4:104–112 (1983).

8. MCKAGUE, E. L., Jr., J. F. Halkias, and J. D. Reynolds. "Moisture Diffusion in Composites: The Effect of Supersonic Service on Diffusion," *J. Composite Materials*, 9:2–9 (1975).

9. BROWNING, C. E., G. E. Husman, and J. M. Whitney. "Moisture Effects in Epoxy Matrix Composites," *Composite Materials: Testing and Design (Fourth Conference)*, ASTM STP 617, American Society for Testing and Materials, Philadelphia, pp. 481–496 (1977).

10. SPRINGER, G. S., ed. *Environmental Effects on Composite Materials*, Technomic Publishing Co., Inc., Lancaster, PA (1981).

11. BUECHE, F. and F. N. Kelley. "Viscosity and Glass Temperature Relations for Polymer-Diluent Systems," *J. Polymer Science*, 45:267–273 (1960).

12. AKLONIS, J. J. and W. J. McKnight. *Introduction to Polymer Viscoelasticity, 2nd ed.* Wiley, New York (1983).

13. SHEN, C. W. and G. S. Springer. "Moisture Absorption and Desorption of Composite Materials," *J. Composite Materials*, 10:2–20 (1976).

14. WHITNEY, J. M. and C. D. Browning. "Some Anomalies Associated with Moisture Diffusion in Epoxy Matrix Composite Materials," *Advanced Composite Materials—Environmental Effects*, ASTM STP 658, Philadelphia, pp. 43–60 (1978).

15. SHIRRELL, C. D. "Diffusion of Water Vapor in Graphite/Epoxy Composites," *Environmental Effects on Advanced Composites*, ASTM STP, American Society for Testing and Materials, Philadelphia (1978).

16. WHITNEY, J. M. and G. E. Husman. "Use of the Flexure Test for Determining Environmental Behavior of Fibrous Composites," *Experimental Mechanics*, to be published.

17. CIRISCIOLI, P. R., W. I. Lee, D. G. Peterson, G. S. Springer, and J. M. Tang. "Accelerated Environmental Testing of Composites," *J. Composite Materials*, 21:225–242 (1987).

18. FOWLER, C. C., Jr. "Bonding of Elevated Temperature Strain Gages to Humid Aged Graphite Tensile Specimens through the Use of Anaerobic Adhesives," *AFML-TR-75-204*, Air Force Materials Laboratory (December 1975).

19. WHITNEY, J. M., C. E. Browning, and A. Mair. "Analysis of the Flexure Test for Laminated Composite Materials," *Composite Materials: Testing and Design (Third Conference)*, ASTM STP 546, American Society for Testing and Materials, Philadelphia, pp. 30–45 (1974).

20. WHITNEY, J. M. *Structural Analysis of Laminated Anisotropic Plates.* Technomic Publishing Co., Inc., Lancaster, PA (1987).

21. WEIBULL, W. "A Statistical Theory of the Strength of Materials," *Ing. Vetenskaps Akad. Handl.* (Sweden), No. 151, pp. 5–45 (1939).

22. HALPIN, J. C., K. L. Jerina, and T. A. Johnson. "Characterization of Composites for the Purpose of Reliability Evaluation," *Analysis of the Test Methods for High Modulus Fibers and Composites*, ASTM STP 521, American Society for Testing and Materials, Philadelphia, pp. 5–64 (1973).

23. KAMINSKI, B. E. "Effect of Specimen Geometry on the Strength of Composite Materials," *Analysis of the Test Methods for High Modulus Fibers and Composites*, ASTM STP 521, American Society for Testing and Materials, Philadelphia, pp. 181–191 (1974).

24. HAHN, H. T. and R. Y. Kim. "Proof Testing of Composite Materials," *J. Composite Materials*, 9:297–311 (July 1975).

25. WEIL, N. A. and I. M. Daniel. "Analysis of Fracture Probabilities in Nonuniformly Stressed Brittle Materials," *J. the American Ceramic Society*, 47:268–274 (June 1964).

26. PAGANO, N. J. and J. C. Halpin. "Influence of End Constraint in the Testing of Anisotropic Bodies," *J. Composite Materials*, 2(1):18–31 (1968).

27. CRON, S. M., H. N. Palazotto, and R. S. Sandhu. "The Improvement of End-Boundary Conditions for Off-Axis Tension Specimen Use," *Experimental Mechanics*, 28:14–19 (March 1988).

28. PINDERA, M. J. and C. T. Herakovich. "Shear Characterization of Unidirectional Composites with the Off-Axis Tension Test," *Experimental Mechanics*, 26:103–112 (March 1986).

29. GRIMES, G. C., P. H. Francis, G. E. Commerford, and G. K. Wolfe. "An Experimental Investigation of the Stress Levels at Which Significant Damage Occurs in Graphite Fiber Plastic Composites," *AFML-TR-72-40*, Air Force Materials Laboratory (May 1972).

30. RYDER, J. T. and E. D. Black. "Compression Testing of Large Gage Length Compression Coupons," *Composite Materials Testing and Design (Fourth Conference)*, ASTM STP 617, Philadelphia, pp. 170–189 (1977).

31. LANTZ, R. B. "Boron Epoxy Laminate Test Methods," *J. Composite Materials*, 3(4):642–650 (October 1969).

32. ADAMS, D. F. and D. E. Walrath. "Current Status of the Iosepescu Shear Test Method," *J. Composite Materials*, 21:494–507 (1987).

33. ROSEN, B. W. "A Simple Procedure for Experimental Determination of the Longitudinal Shear Modulus of Unidirectional Composites," *J. Composite Materials*, 6(4):552–554 (October 1972).

34. LIFSHITZ, J. M. "Determination of Nonlinear Shear Modulus of a Fiber-Reinforced Lamina from the Axial Behavior of [±45], Specimens," *J. Composite Technology and Research*, 10:146–150 (1988).

35. CHAMIS, C. C. and J. H. Sinclair. "Ten-Degree Off-Axis Test for Shear Properties in Fiber Composites," *Experimental Mechanics*, 17(9):339–346 (September 1977).

36. HAHN, N. T. and J. Erikson. "Characterization of Composite Laminates Using Tubular Specimens," *AFML-TR-77-144*, Air Force Materials Laboratory (August 1977).

37. PAGANO, N. J. "Stress Gradients in Laminated Composite Cylinders," *J. Composite Materials*, 5:260–265 (April 1971).

38. PIPES, R. B., D. L. Reed, and J. E. Ashton. "Experimental Investigation of the Interlaminar Shear Properties of Composite Materials," Spring·Meeting and Exposition, Society for Experimental Stress Analysis, Cleveland, Ohio (1972).

39. BERG, C. A., J. Tirosh, and M. Israeli. "Analysis of Short Beam Bending of Fiber Reinforced Composites," *Composite Materials Testing and Design (Second Conference)*, ASTM STP 497, Philadelphia, pp. 206–218 (1972).

40. PIPES, R. B. and B. W. Cole. "On the Off-Axis Strength Test for Anisotropic Materials," *J. Composite Materials*, 17:246–256 (1973).

41. WHITNEY, J. M., N. J. Pagano, and R. B. Pipes. "Design and Fabrication of Tubular Specimens for Composite Characterization," *Composite Materials Testing and Design (Second Conference)*, ASTM STP 497, Philadelphia, pp. 52–67 (1971).

42. PAGANO, N. J., J. C. Halpin, and J. M. Whitney. "Tension Buckling of Anisotropic Cylinders," *J. Composite Materials*, 2:154–167 (April 1968).

43. WHITNEY, J. M. and C. T. Sun. "A Refined Theory for Laminated Anisotropic Cylindrical Shells," *J. Applied Mechanics*, 41:471–476 (June 1974).

44. DANIEL, I. M. "Biaxial Testing of Graphite/Epoxy Composites Containing Stress Concentration," *AFML-TR-76-244, Part I*, Air Force Materials Laboratory (December 1976).

45. DANIEL, I. M. "Biaxial Testing of Graphite/Epoxy Composites Containing Stress Concentrations," *AFML-TR-76-244, Part II*, Air Force Materials Laboratory (June 1977).

Ultrasonic Nondestructive Evaluation Techniques for Composite Materials

R. A. BLAKE, JR.

Ultrasonic nondestructive inspection techniques are useful in both quality control and research applications for flaw detection in fiber reinforced composite materials. The technique can be used to characterize material flaws such as delaminations, cracks, voids, and matrix rich pockets; to visualize fiber orientation; to measure changes in thickness; and to determine material elastic properties. For homogenous materials, like metals, ultrasonic techniques are well developed and reasonably simple. The nonhomogenous nature of composites presents an added dimension of complexity to ultrasonic inspection techniques. The principal problems are reproducibility of results and flaw discrimination (the correlation of characteristic ultrasonic data to a specific type of flaw or material micro-structure).

Ultrasonic material inspection is accomplished by generating sound energy (at ultrasonic frequencies) and directionally focusing the energy on the material to be inspected. The sound source is coupled to the material under inspection by water because of water's efficiency in transmitting low power sound energy. The ultrasound energy travels through the material under inspection, and a complex interaction takes place with the material micro-structure. Energy is reflected from interfaces where the material density changes. It is the measurement of the changes in the ultrasonic energy resulting from its interaction with the material under test that can be correlated to information about the material structure. The techniques used to measure and interpret the ultrasonic information constitute the major task involved in ultrasonic NDE. The methods employed to accomplish this task have profound effects on resolution in flaw detection.

In this section an attempt is made to describe the equipment and techniques used to characterize flaws in composite materials by the ultrasonic method. The methods described herein constitute state-of-the-art developments in an evolving technology; therefore, emphasis has been placed on the basic principles and their application in the detection of damage or flaws. In the Appendix, selected examples of composite material C-scans are presented.

6.2.1 Ultrasonic Inspection System Description

Basic System (Analog)

The primary functions performed by a basic ultrasonic system are generation of an ultrasonic pulse of energy, coupling (through water) of the ultrasound to the material under test, reception of the ultrasound after it has interrogated the material, analysis of this ultrasonic information, and display of the analyzed information. The boxed region of Figure 6.2-1 shows the basic components that perform these tasks. In the example system shown, ultrasonic signals originate from a Parametrics 5052 UA ultrasonic analyzer that excites a lithium sulphate transducer to produce sound in the 1 to 20 MHz range. The same transducer is used to receive the sound after it has propagated through the material under test and the water coupling medium. The ultrasonic analyzer is used to select the portion of the incoming ultrasonic waveform to be analyzed (gating), and the gated portion of the waveform is subjected to peak height analysis wherein the peak output is quantized to ten discrete levels in 0.1 volt increments. Each level corresponds to a distinct shade of gray that increases in darkness with increasing voltage level. The output is transmitted to a pen amplifier where a signal is produced that forms an image by burning the surface layers of an ink-impregnated electrically conductive paper. In order to inspect the whole surface, the transducer must be moved over each point of the material. The resultant two-dimensional gray scale image produced from a point-by-point scan of the surface is called a C-scan.

It should be mentioned that ultrasonic C-scans are usually made by immersing the part under inspection in a water bath. The transducer is placed into the water bath on the end of a search tube attached to a manipulator. Most mechanical scanning systems allow only planar scanning motions of flat or circular symmetric shapes. Water jet techniques have been developed that allow inspection of parts too large to be immersed in a tank. Current technology exists for robotically controlled manipulators that can track complex surface geometries, but such systems are not yet commonplace. If a focused transducer is used, the distance between the transducer and test piece must be held constant. The angle between the axis of the transducer and the normal to the surface of the test piece is determined by the type of ultrasonic test conducted. Since the speed of the test is limited primarily by the speed of the scanning, arrayed transducers are often used for large-scale inspection tasks to reduce scan times.

Digital System

Analog analysis methods have limitations that can be overcome by the introduction of digital techniques. The remaining portion of Figure 6.2-1 shows an example of the extended ultrasonic system capable of digital ultrasonic analysis. The digital approach increases the power and versatility of the inspection system.

There are many techniques that can be implemented digitally to improve and quantify the ultrasonic inspection process. Methods are given later for the elimination of some of the undesirable qualities attached to analog scanning and image formation. The Fourier transform is used for spectral analysis and for echo peak enhancement that precisely locates flaw depth. Velocity-of-sound measurements may be made as a function of frequency using phase information obtained from the Fourier transform.

The signature formed by digitally monitoring the amplitude of a specified segment of the total ultrasonic waveform has attributes that can be related to particular material discontinuities. The time domain representations of the ultrasonic waveform can then be used to distinguish between discontinuities that appear similar in the amplitude signature.

Digital imaging is used to obtain control of display parameters and results in superior image reproducibility, improved signal-to-noise ratios, image magnification, and decreased image formation times. Digital image formation and storage allows the area image processing to be incorporated in the ultrasonic nondestructive evaluation of composites. Image histogram and two-dimensional Fourier transform techniques are expected to significantly improve image quality and feature extraction. Finally, the addition of color graphics has been demonstrated to be an aid in the interpretation of ultrasonic data.

In the digital approach the ultrasonic analyzer is used to send and receive ultrasonic pulses. The incoming signal is digitized at a 20 MHz rate by an oscilloscope after prefiltering to band-limit the signal to 10 MHz and prevent aliasing. The digitized waveform is then fed to a computer for signal processing. Peak data from multiple gates are deciphered by a separate microprocessor to pipeline the data flow and increase interrogation rates. The microprocessor performs gray scale conversions and outputs these data to either the dot matrix printer, the analog conductive paper system, or the color graphics system.

The dot matrix printer produces ordered dither images of the C-scan information and has an image density of 19 by 22 dots per cm^2 (120 by 140 dots per in^2). The images are formed on standard tractor feed computer paper.

The video graphics computer allows for the display of color C-scans in a variety of formats. The unit has a 512×512 pixel resolution and displays 16 colors simultaneously that are selected from a color lookup table containing more than 16 million possible shades.

Fourier analysis and flat bed plotter graphics are done by a general purpose central computer system after transmitting the digitized waveform over a serial line that connects the systems.

6.2.2 Basic Types of Scans

There are three basic types of scans referred to in the field of ultrasonic nondestructive evaluation: the A-scan, B-scan, and C-scan. The distinctions among the three lie in the interpretation and representation of the ultrasonic data. Descriptions of each method are discussed next along with the pros and cons with respect to flaw detection and characterization.

A-Scan

The A-scan in Figure 6.2-2 is the time domain representation of an ultrasonic waveform produced by sound

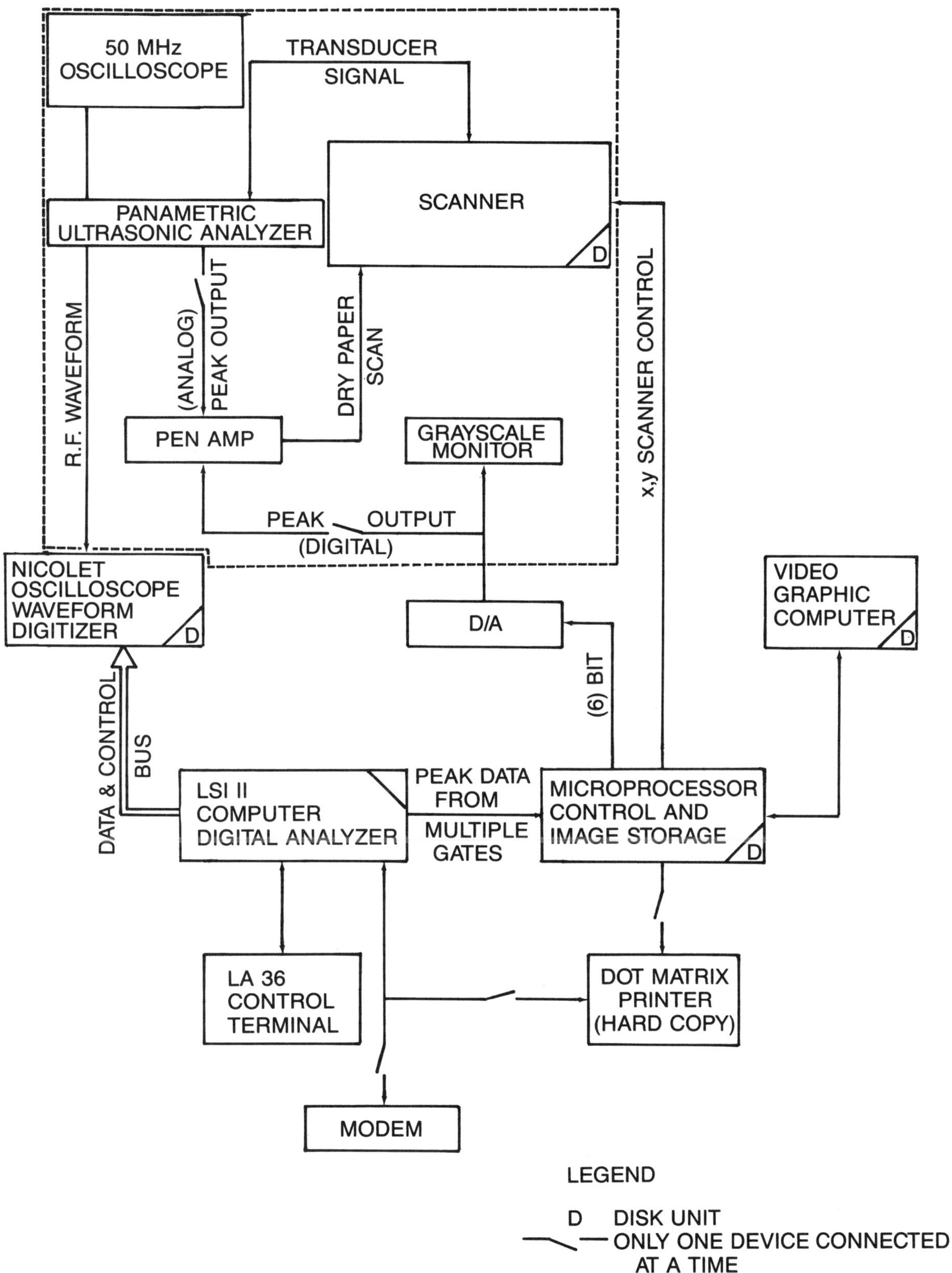

FIGURE 6.2-1. Block diagram example of digital ultrasonic inspection system.

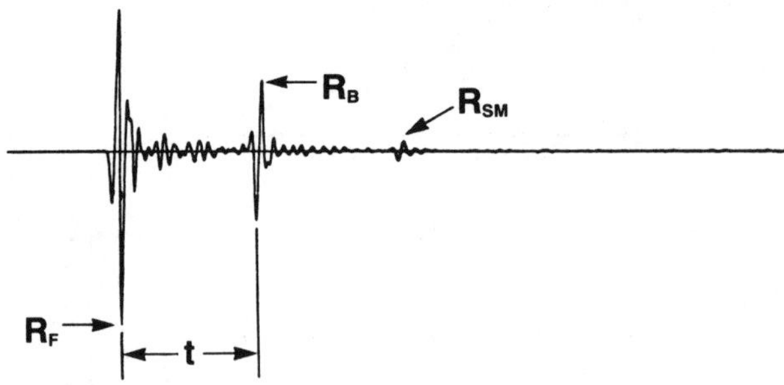

FIGURE 6.2-2. Ultrasonic A-scan depicting velocity of sound and attenuation coefficient parameters.

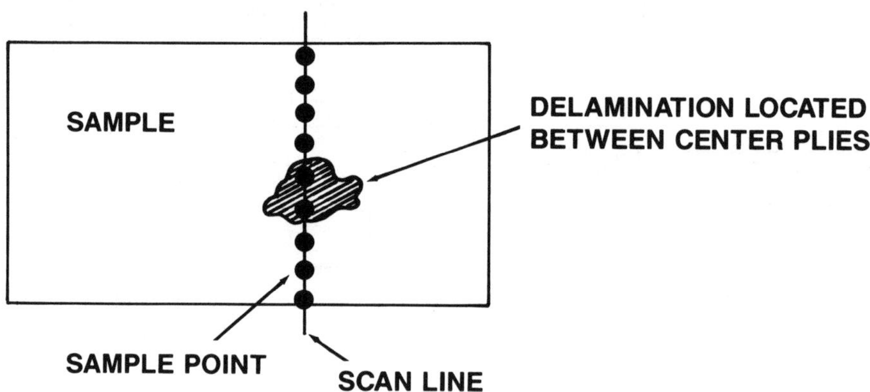

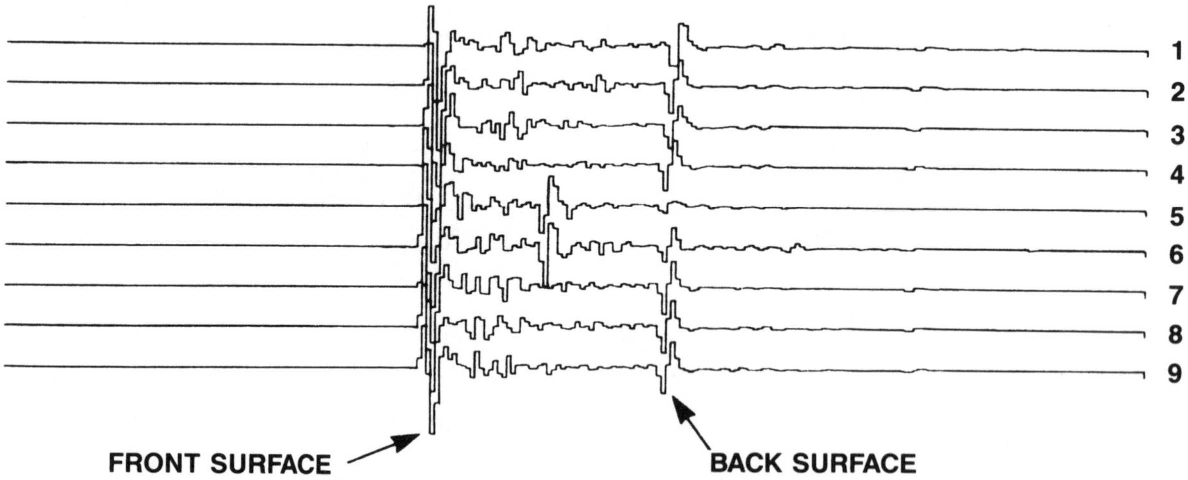

FIGURE 6.2-3. Representation of a delamination by a family of A-scans.

in the 2.5 to 30 MHz frequency range reflecting from the front and back surfaces of the material under inspection. Figure 6.2-2 shows an A-scan of a reflected ultrasonic waveform with the front surface reflection and the back surface reflection labeled R_F and R_B respectively. Also noted is the second multiple echo from the sonic interrogation, labeled R_{SM}. Interactions with unidirectional or woven plies, delaminations, voids, inclusions, and matrix rich regions all produce reflections that occur between the front and back surface echoes. It is the evaluation of these reflections and correlation of the results to specific types of damage or material structure that forms the basis of quantitative ultrasonic NDE.

The distinguishability among reflections from different sources is a complex issue that will be discussed in detail in the segment on amplitude signature evaluation; only a brief description is pertinent here. Basically three types of information are contained in the A-scan: time, amplitude, and frequency. This information must be used to locate and differentiate specific types of flaws and material structure.

The reflections in composite materials formed from unidirectional prepreg tape are low in amplitude as compared with those of woven material systems. Woven materials produce large amplitude periodic reflections that decay logarithmically and terminate at the back surface reflection. These periodic reflections are produced by the matrix pockets formed in the interstitial spaces of the weave; they oscillate with a period proportional to the weave spacing.

Matrix rich regions between the plies of a composite material produce interlaminar echoes that attenuate the back surface reflection but do not eliminate it. Delaminations totally eliminate the back surface reflection and produce an echo as large as possible with the available energy. Second and third multiple reflections of the delamination are produced at regular intervals, provided the delamination is open and sufficient energy is reflected by the interface.

Several geometric factors affect the flaw resolution attainable using ultrasonic inspection. Void geometry is an important consideration because spherical and cylindrical voids allow sound to propagate around them, in addition to producing a reflection corresponding to the depth of the void. The position of the void with respect to the focal plane of the transducer also affects the response. The combined effects of void geometry, location, and size may allow a back surface reflection to be

produced in addition to the reflection from the void. An example of this effect for small diameter edge drilled holes will be presented later in the discussion on digital systems.

Velocity of sound and the linear attenuation coefficient may be calculated directly from the A-scan or evaluated using Fourier analysis. The velocity of sound, C, is measured perpendicular to the plies of the composite and provides an acoustic measure of the elastic constant, C_{33}, according to Equation (6.2-1):

$$C_{33} = \varrho c^2 \qquad (6.2\text{-}1)$$

where

$c = 2d/t$
$d = $ laminate thickness
$t = $ interface delay time

The linear amplitude attenuation coefficient, α, is determined from measurements of the amplitudes of the front surface, back surface, and second multiple echoes using Equation (6.2-2):

$$\alpha = \frac{-1}{2d}\, \ell n\, [(R_{sm}/R_B) - (R_B/R_F)] \qquad (6.2\text{-}2)$$

where

$R_F = $ amplitude of front surface reflection
$R_B = $ amplitude of back surface reflection
$R_{sm} = $ amplitude of second multiple echo

Attenuation is a useful parameter for characterizing various types of damage in composite materials.

A family of A-scans recorded at equal spatial distances displays an image of a plane perpendicular to the plies of the composite. Figure 6.2-3 demonstrates the representation of a delamination by a family of A-scans. The A-scan family in this example was formed by the digital time sequential sampling of the ultrasonic waveform at a 20 MHz rate. The sample points were connected with straight lines, which results in a stepped waveform.[1] The delamination shows up as a reflection located between the front and back surface echoes of scan lines 5 and 6. The information derived from a family of A-scans can be analyzed using various techniques, which are discussed later, to form two-dimensional images of material structure.

---

[1]The section on Fourier analysis discusses an interpolation technique that smooths the digital A-scan representation.

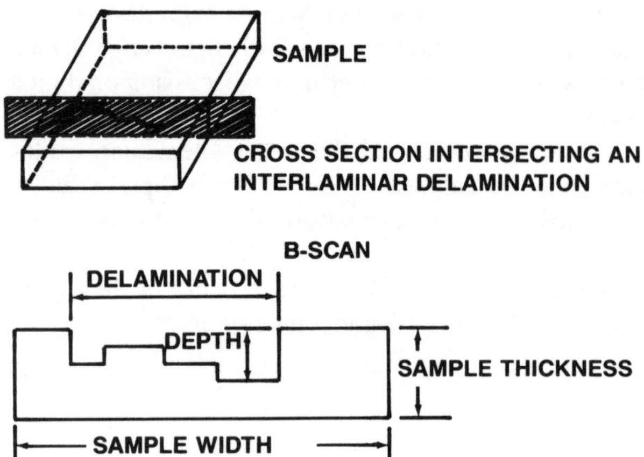

FIGURE 6.2-4. B-scan representation of delaminated region.

B-Scan

The B-scan shown in Figure 6.2-4 gives flaw depth and thickness information for a single cross-sectional plane normal to the sample. B-scans are formed from the time-of-flight measurement calculated from the time between the front surface echo and the occurrence of the next echo above a set threshold. The B-scan can be displayed directly, or multiple B-scans may be stored and manipulated to form a C-scan on a time basis, as discussed in the next segment.

The threshold mentioned previously may be a constant or a function of time that compensates for the logarithmic attenuation of sound in the material. The subsection on digital systems discusses threshold compensation for multigate systems.

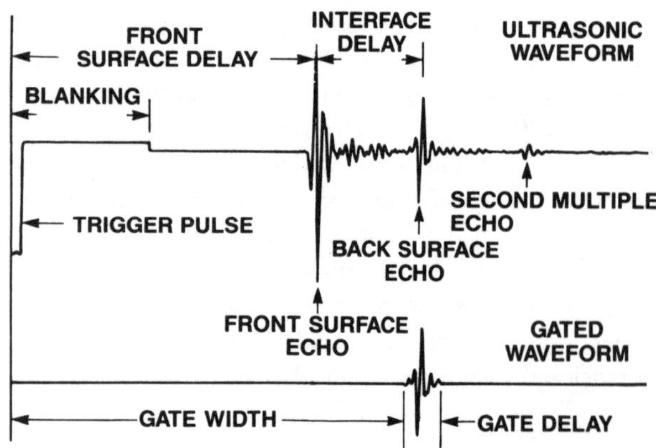

FIGURE 6.2-5. C-scan gate parameters.

Conventional B-scans indicate only the occurrence of the first echo after the front surface reflection; they are incapable of displaying second and higher multiple reflections that distinguish voids from matrix rich regions. Sophisticated computer-based ultrasonic systems produce B-scan representations that image information as it is received by the transducer. These systems display images based upon density changes in addition to the reflection and scattering of sound in the material. Sector scans formed in this manner have found a wide range of medical diagnostic applications; similar methods can be used for the nondestructive evaluation of composite material systems.

Isometric projections may be formed from B-scan data by adding z-axis information to the display. The image is formed on a cathode ray tube (CRT) or computer graphic display, and the ability to tilt, rotate, and magnify the image facilitates interpretation.

C-Scan and Gating Methods

C-scans are two-dimensional, in-plane views of the internal structure of a material. The image is created from either peak height or time-of-flight data; it is displayed on computer peripherals or electrically conductive paper. C-scans may also be formed from frequency domain data obtained through Fourier analysis of the ultrasonic waveform.

The gated region in Figure 6.2-5 is a specified segment of the total ultrasonic waveform that is analyzed on either a peak amplitude or spectral basis to form a C-scan. The interface delay is a function of the velocity of sound and thickness of the material, and the front surface delay is dependent upon the focal length of the transducer. For long focal length transducers, sound echoes from a previous pulse or event may occur prior to the front surface echo, in which case blanking may be used to prohibit false triggering of the gate delay. The gate delay may be triggered from either the trigger pulse or the front surface echo, but preferably from the latter in order to compensate for warpage and slight thickness irregularities.

The gate width is set to include as much of the waveform as necessary for the type of scan desired. As discussed later in this segment, the specification of the gate parameters dominates the results of the C-scan interrogation. Accurate methods of recording these data are mandatory for insurance of image reproducibility.

X1

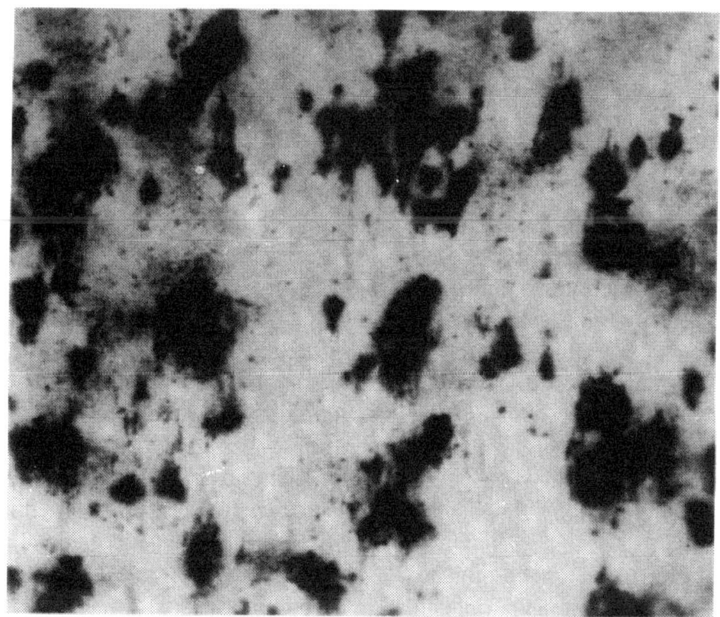

X100

FIGURE 6.2-6. Conductive paper C-scan recording.

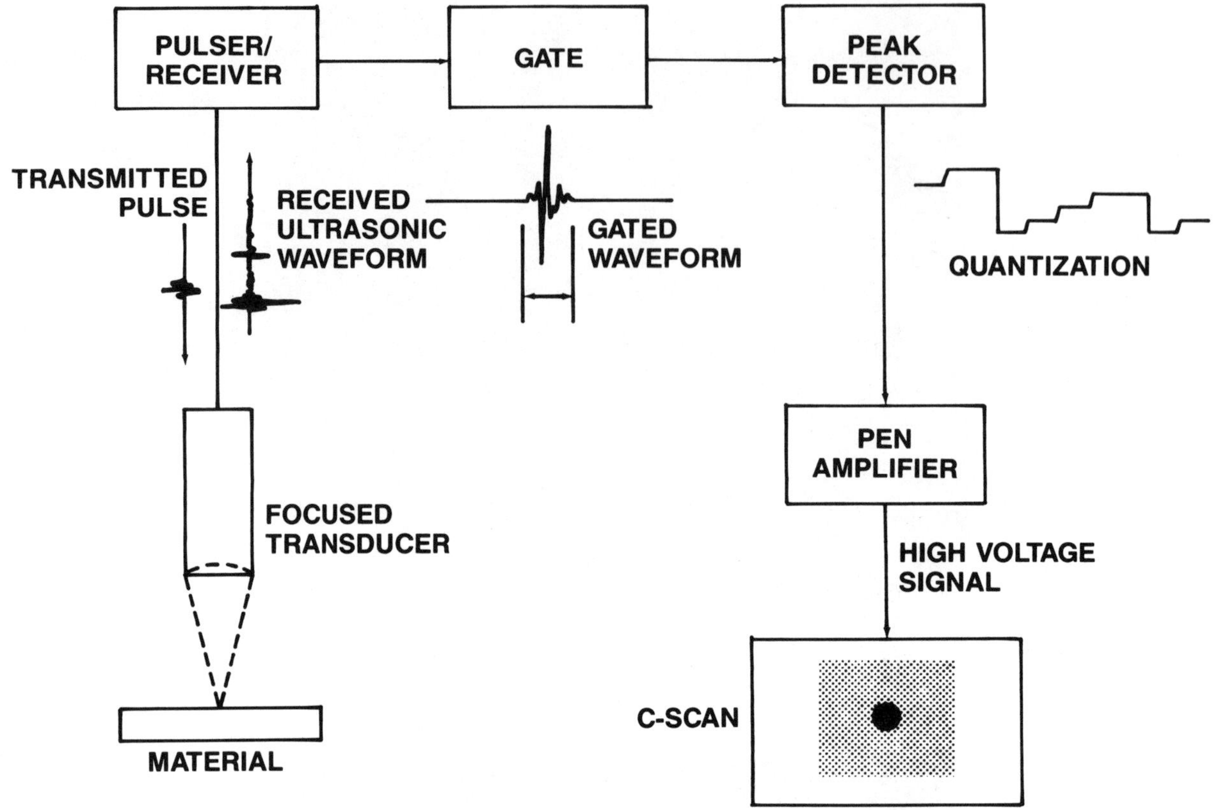

FIGURE 6.2-7. Pulse-echo C-scan by peak amplitude analysis.

Gate selection determines whether the C-scan will be based upon attenuation or echo capture data. Echo characterization, which also depends upon gate selection, will be discussed in the subsection on delamination characterization. Combined methods and multigate techniques are useful, although data interpretation becomes more complicated. In general, the C-scan will indicate the location of a discontinuity in two or three dimensions; however, until more sophisticated ultrasonic adaptive analyzers and color display systems are combined, the A-scan must be evaluated to determine the identity of the flaw.

Ultrasonic C-scan formation has been accomplished using a variety of methods. The three primary techniques will be outlined for amplitude, interface, and spectral analysis. The display medium in each example is assumed to be an electrically conductive paper that records an image by burning discrete quantities of the surface coating from the paper as shown in Figure 6.2-6. Variability in the display medium, as well as

computer graphic displays, will be discussed in the subsection on digital image processing.

The method of C-scan formation shown in Figure 6.2-7 is based upon peak amplitude analysis of the gated signal. A large-amplitude, short-duration pulse is transmitted to a lithium sulphate transducer that produces sound in the 2.5 to 20 MHz frequency range depending upon the transducer element. The echoes reflected from the material are received by the same transducer in the pulse-echo mode or by a second transducer in the pulse-catch or through-transmission mode; a gated region is analyzed for positive or negative peak amplitude. The amplitude is quantized to ten discrete levels corresponding to monotonically increasing shades of gray on the C-scan. Earlier C-scan systems displayed images on binary displays such as a CRT monitor but allowed for only one threshold setting instead of a continuous or discrete gray scale. These systems rely upon reference specimens for threshold adjustment and produce extremely poor results that preclude accurate interpretation. Slight adjustment of the threshold setting produces totally different results, causing unacceptable image reproducibility. Binary displays can be used to form gray scale C-scans, and a discussion of these techniques will be given in the subsection on digital image processing.

The interface delay may be measured and converted into a gray scale value on a C-scan image as diagrammed in Figure 6.2-8. A base level for a constant thickness material is set, and the apparent thickness variation due to the presence of a delamination produces a corresponding change in the C-scan intensity. The C-scan in this instance is based upon the B-scan rather than the A-scan signal. Systems that employ this method of C-scan formation are best suited for weld and corrosion inspections in metals and require additional signal processing to reliably inspect composite materials. This restriction is due to the complexity of the composite material waveform as compared to an isotropic, homogeneous material system.

The C-scan may be formed from a selected frequency component of the gated ultrasonic waveform by Fourier transforming the ultrasonic signal in an analog spectrum analyzer as indicated in Figure 6.2-9. This method of C-scan formation requires more time than the two previously mentioned methods due to the conversion rate imposed by the spectrum analyzer. If the scan speed is increased, hysteresis and a low pass filtered image will result. Low pass filtering causes the image to appear out of focus, and hysteresis produces a double image. The frequency-based C-scan allows for the maximum control of resolution and frequency content; however, the transducer selection is more important here than with the other two methods, since the frequency response of the transducer directly affects the image quality. Phase-insensitive power transducers appear best suited for this analysis.

The gating techniques previously discussed mention attenuation, echo capture, echo characterization, and combined methods. Examples of these gating forms will now be presented for amplitude-based pulse-echo C-scans.

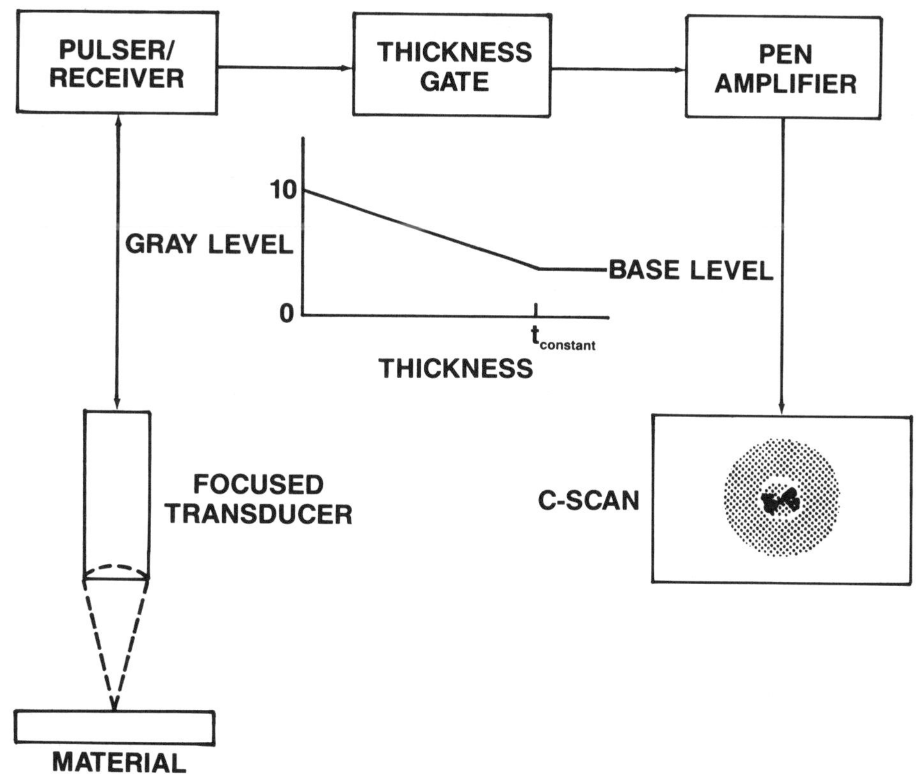

FIGURE 6.2-8. Pulse-echo C-scan using interface delay measurements.

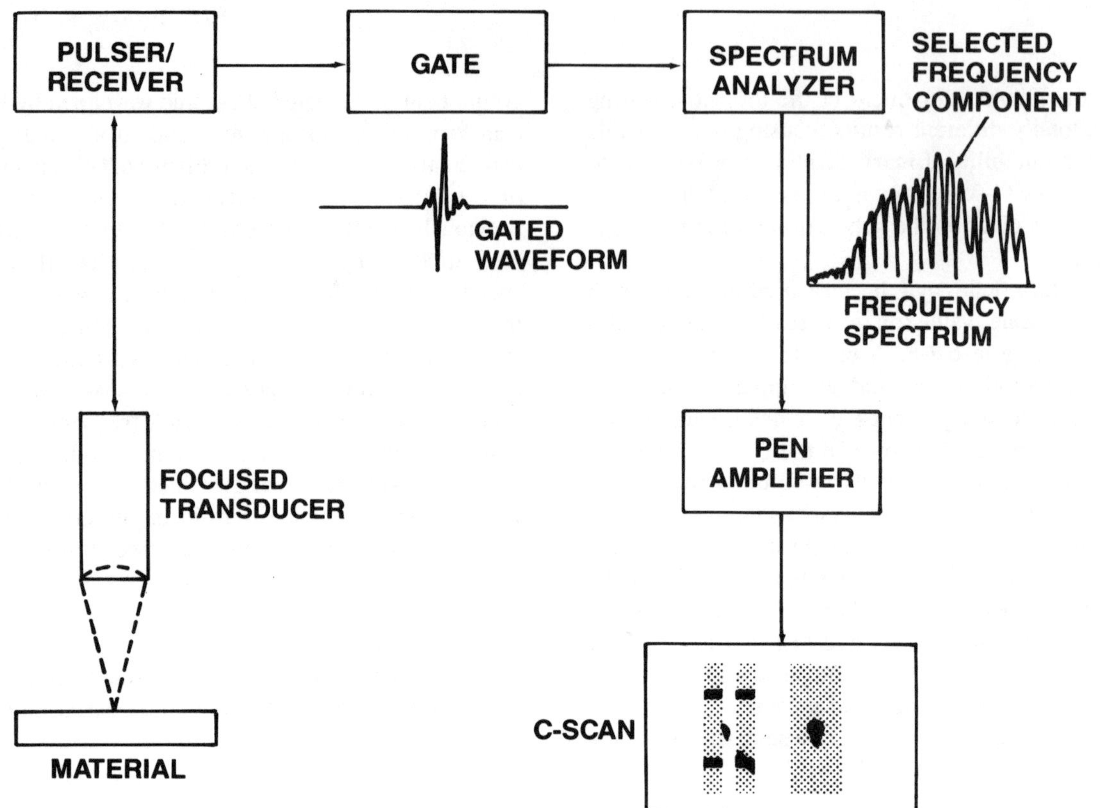

FIGURE 6.2-9. Pulse-echo C-scan by spectral analysis.

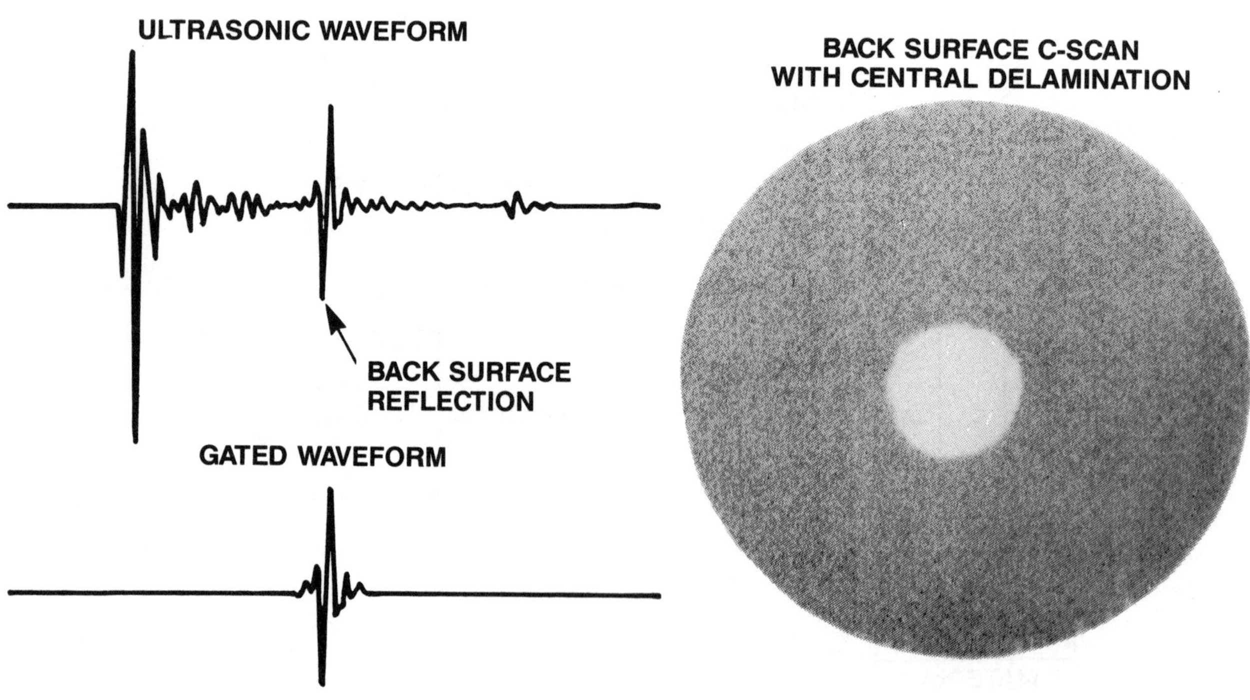

FIGURE 6.2-10. Attenuation dominated C-scan.

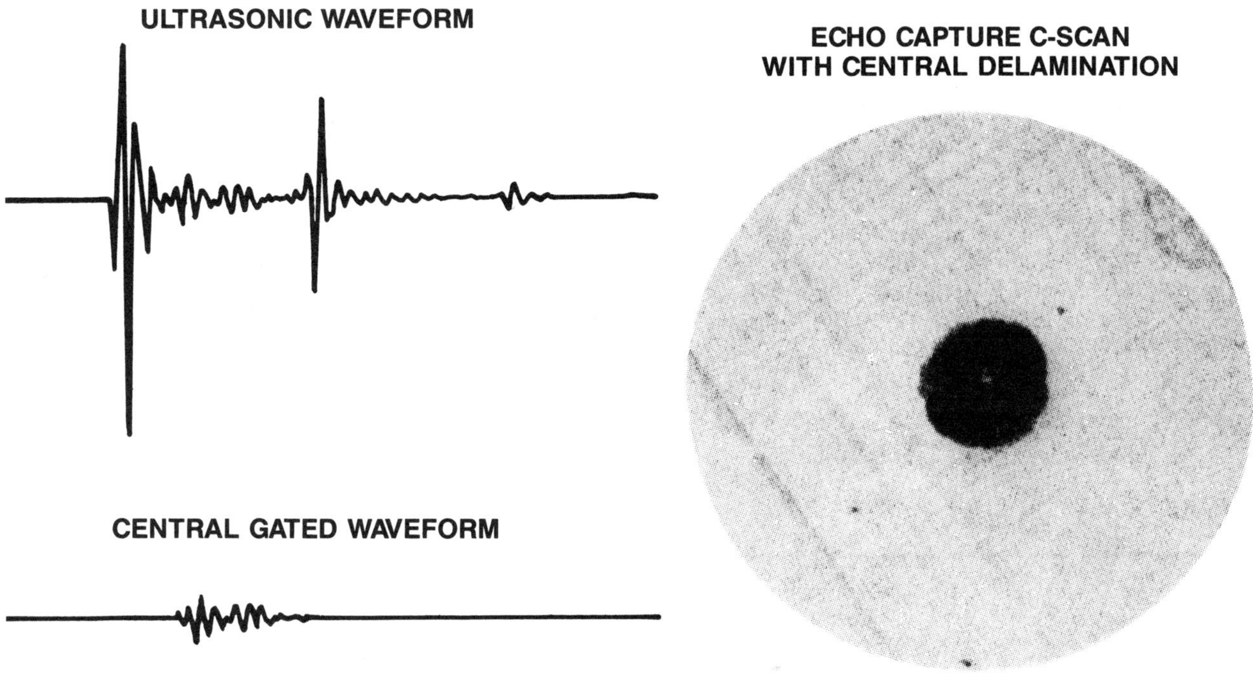

FIGURE 6.2-11. Echo capture gating results.

A C-scan formed with the gated region centered on the back surface produces the attenuation dominated image shown in Figure 6.2-10. The presence of a discontinuity will appear lighter than the surrounding region, since energy was reflected by the flaw and attenuates the amplitude of the back surface reflection.

If the gated region is set to analyze information from the central portion of the ultrasonic waveform, exclusive of the front and back surface reflections, a C-scan will be formed that emphasizes the presence of interlaminar echoes. This result is shown in Figure 6.2-11 for a disk with a central delamination.

The gated waveform may be specified to examine a narrow segment of the total waveform that characterizes the response from a two-ply thick region of the material. Figure 6.2-12 shows that fiber orientation and interlaminar echoes from between the two plies will be emphasized.

If the gated region combines the previous methods to include the back surface reflection and a substantial portion of the central region, the C-scan will indicate delaminations near the front surface as light regions due to attenuation, and delaminations near the back surface will appear dark due to echo capture. The complex central delamination in Figure 6.2-13 demonstrates this result. Caution should be exercised, since no indication will be produced if the delamination echo is the same amplitude as the back surface reflection. In general, back surface gating is used to determine the extent of damage due to impact, fatigue, or static loading; the combined gating method is used as a prelude to echo characterization.

The previous discussion has assumed normal incidence with regard to the transducer and the plane of the material under inspection. If the transducer is set at an angle, θ, to the material, the presence of cracks through the plies of the composite becomes detectable. This type of C-scan is referred to as a *dark field* or a *backscatter* C-scan. Small surface irregularities are minimized, and only reflections from structures inside the material are returned to the transducer. The angle of refraction will not be a straight line as the representation of Figure 6.2-14 indicates but will depend upon the material properties.

The dominating factors for this type of C-scan are scan direction, transducer angle, surface roughness,

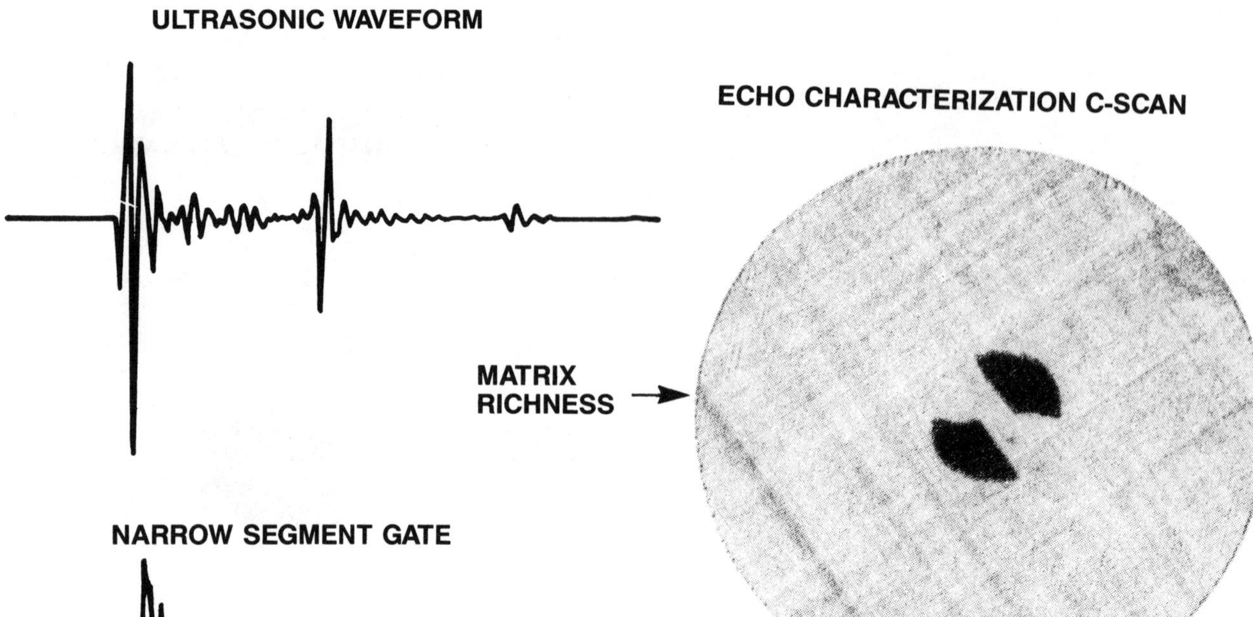

FIGURE 6.2-12. Echo characterization example.

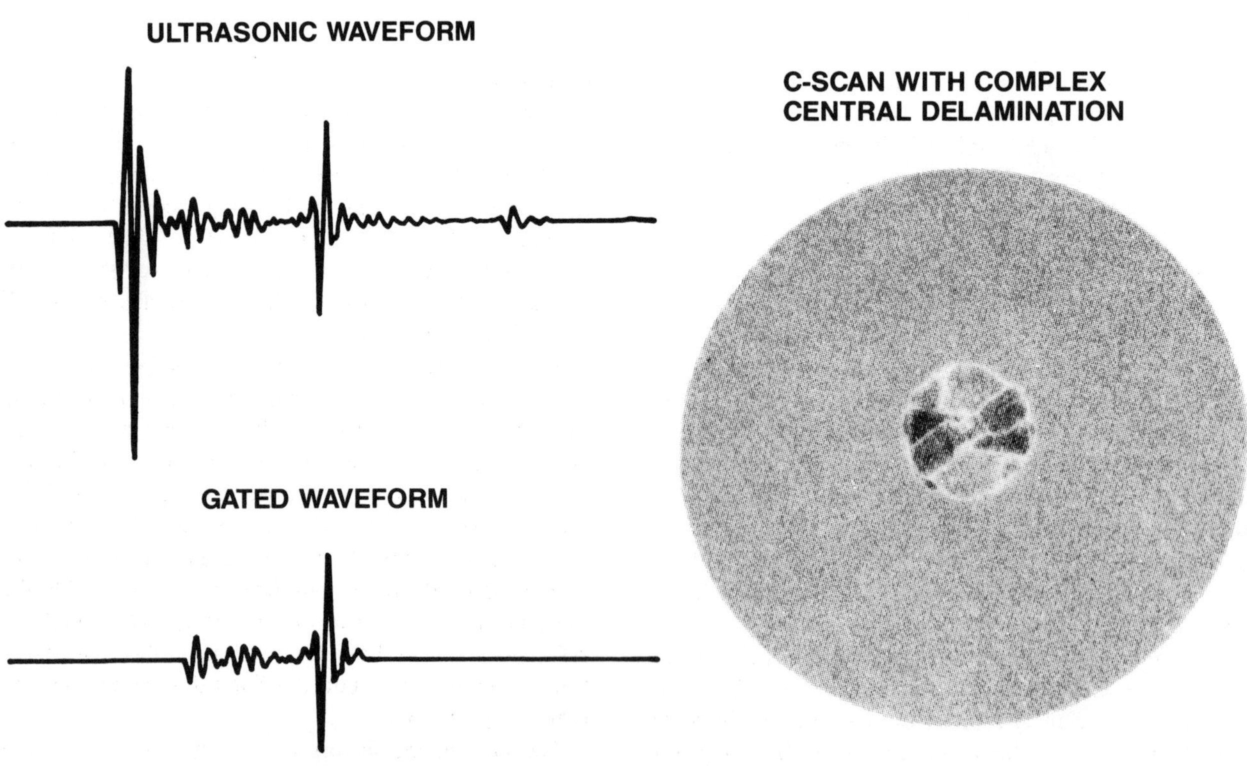

FIGURE 6.2-13. Combined gating method.

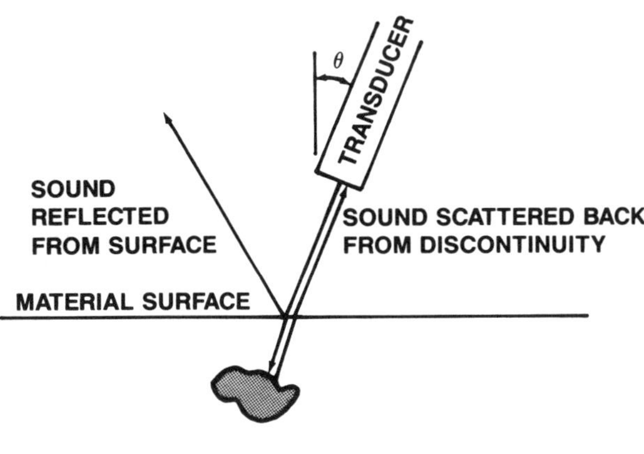

FIGURE 6.2-14. Backscatter/darkfield representations.

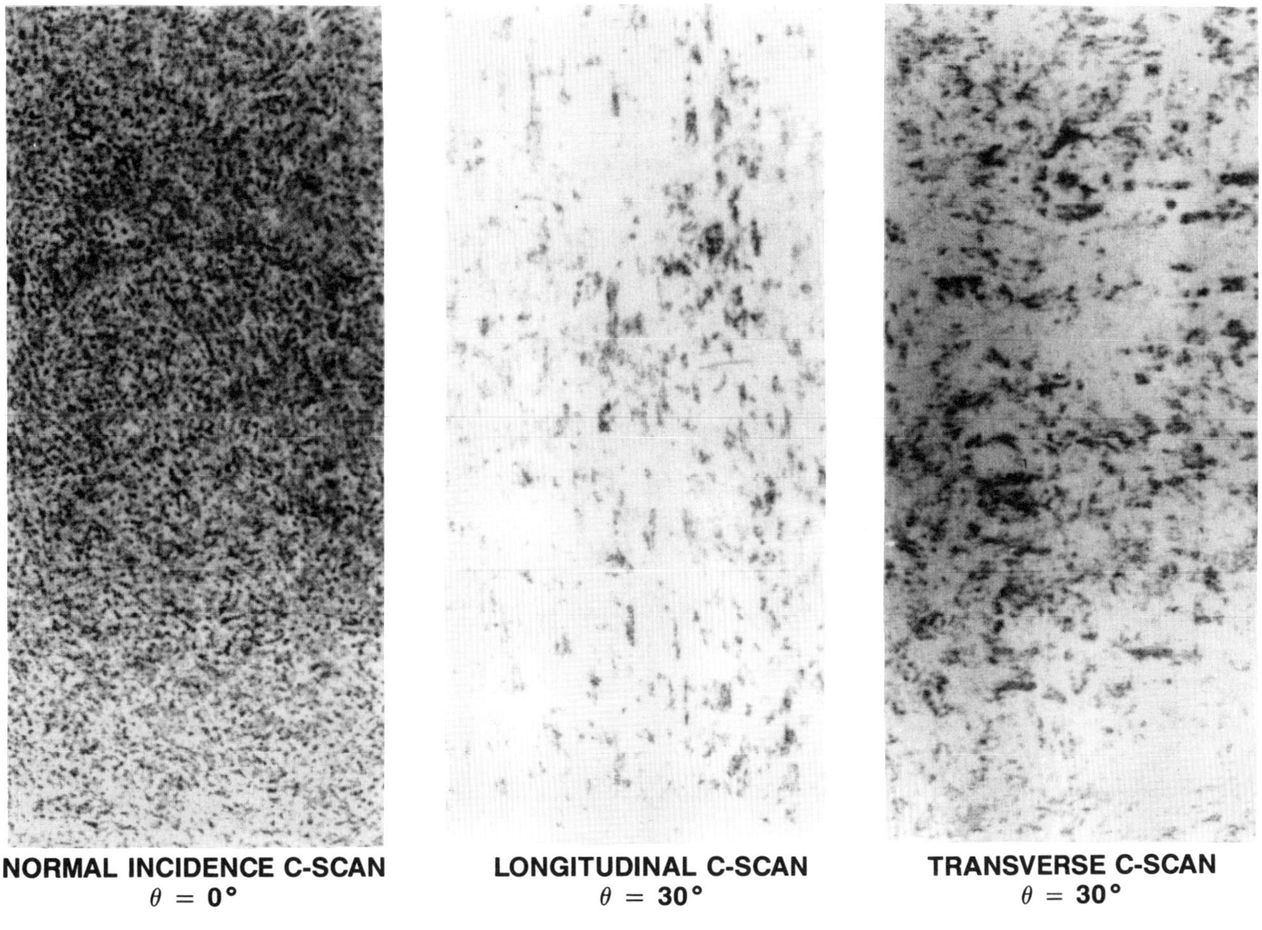

NORMAL INCIDENCE C-SCAN
$\theta = 0°$

LONGITUDINAL C-SCAN
$\theta = 30°$

TRANSVERSE C-SCAN
$\theta = 30°$

FIGURE 6.2-15. Backscatter/darkfield C-scan examples.

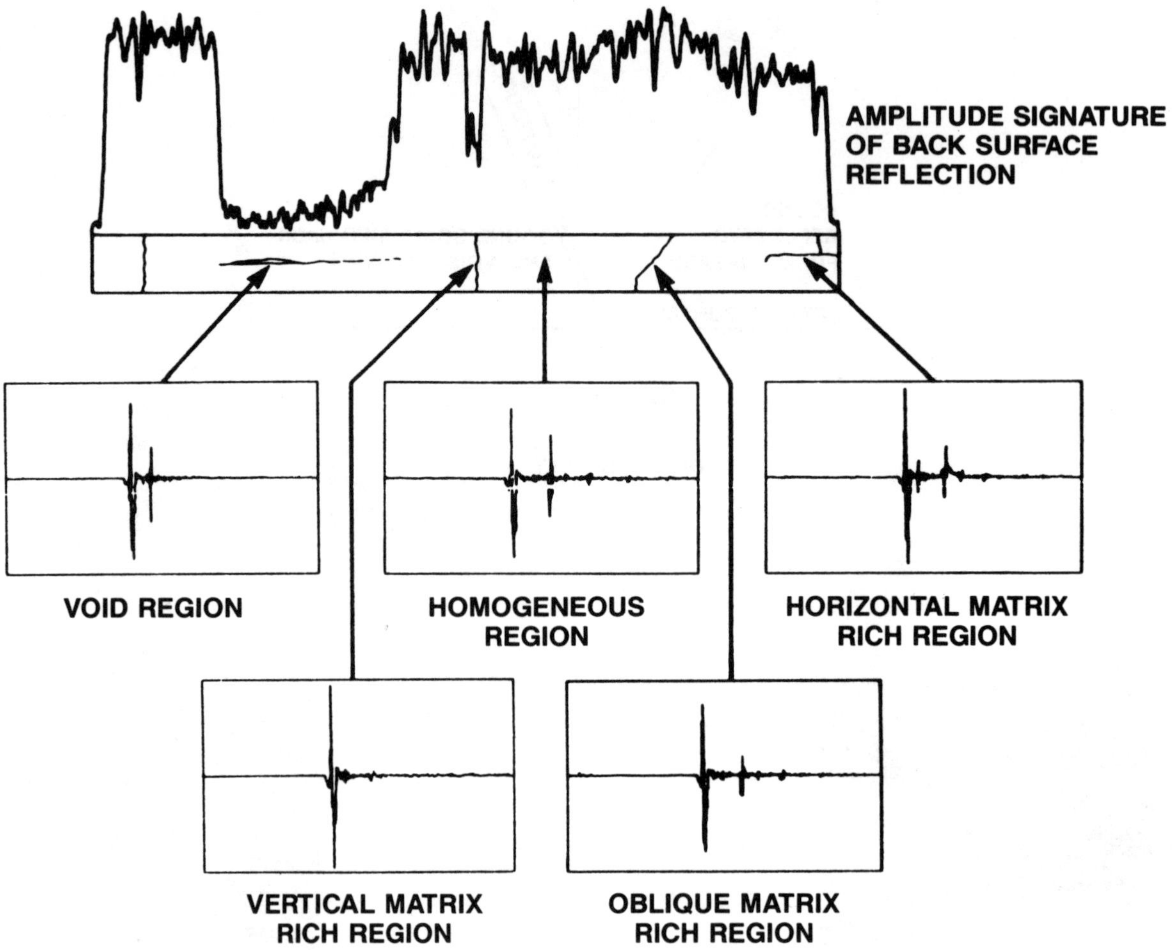

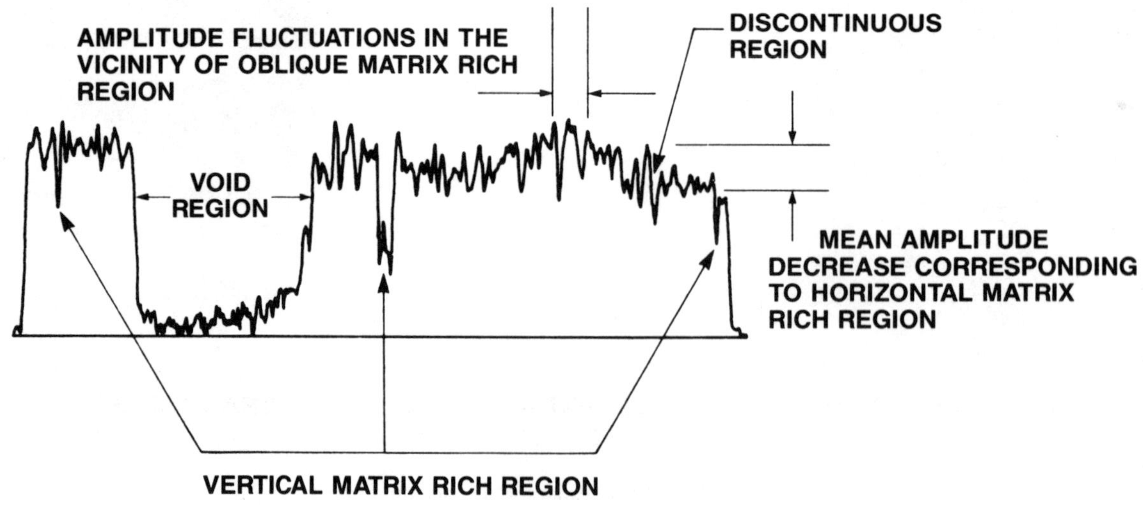

FIGURE 6.2-16. Amplitude signature and time domain waveforms for unidirectional composite materials.

FIGURE 6.2-17. Evaluation of back surface reflection amplitude signature.

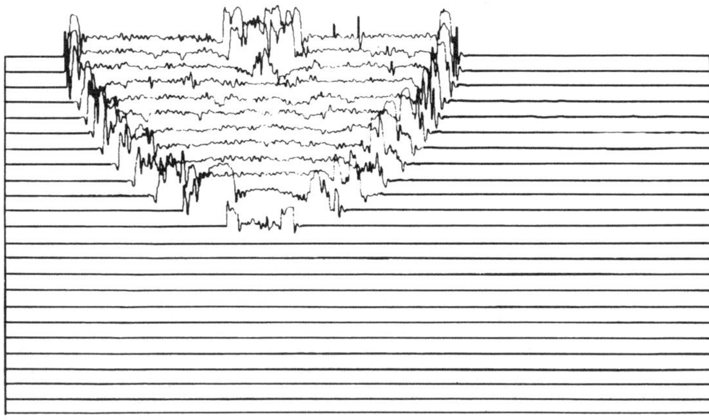

FIGURE 6.2-18. Family of amplitude signatures for semicircular disk with central disbond and delaminated circumference.

and material thickness. Of these factors, scan direction is of primary importance since a reflection will be produced only when the crack or fiber direction is normal to the transducer and in line with the scan direction. The longitudinal and transverse backscatter C-scans in Figure 6.2-15 demonstrate this result for a transducer angle of 30°. The material under inspection is a 65% weight fraction SMC with a semi-smooth surface. In this instance, surface effects dominate the normal incidence C-scan and are reduced through backscatter C-scanning. The backscatter images do reveal internal structure; however, a complete inspection of this type for 360° rotation about a point must be automated—it would be extremely time consuming and difficult to display and interpret.

6.2.3 Amplitude Signature Evaluation

The three primary discontinuities associated with composite materials are voids, inclusions, and matrix richness. Each of these discontinuities affects the ultrasonic response differently, as indicated by Figure 6.2-16. The figure shows an amplitude signature formed by the digital time sequential sampling of the peak response output from the back surface reflection as the sample is scanned at constant velocity. The identifiable characteristics of the amplitude signature are labeled in Figure 6.2-17.

As the ultrasonic waveforms in Figure 6.2-16 indicate, the back surface reflection is severely attenuated in vertical matrix regions, and interlaminar activity is slightly higher than the homogeneous region. Large amplitude reflections are not present in the vertical

matrix regions; thus, a vertical region may be distinguished from a horizontal region or void. The back surface attenuation associated with a horizontal matrix rich region is accompanied by an interlaminar echo whose amplitude is significantly above the background level and whose position is directly related to the depth of the region. The interlaminar echo associated with the horizontal matrix rich region may be distinguished from that of a void region by the presence of the back surface reflection at the proper location. A void region produces a large echo that totally attenuates the back surface reflection and produces equal distant second multiple echoes of the void, provided the initial interlaminar echo is of sufficient amplitude. The response of the ultrasonic wave to an oblique matrix rich region is similar to that of a horizontal matrix rich region; therefore, it is difficult to distinguish. Both types of regions are characterized by an attenuation of the back surface reflection and an increase in interlaminar activity. Presumably, an oblique matrix rich region could be distinguished from a discontinuous region through interpretation of shear wave data rather than compressional wave information. The discontinuous region identified in Figure 6.2-17 was found to contain inclusions and a lower fiber volume fraction, as verified by sectioning and microscopic examination.

If the amplitude signature is formed from interlaminar data rather than attenuation data from the back surface reflection, the result is a signature which emphasizes the presence of voids. A family of such signatures may then be grouped spatially to form a pseudo three-dimensional image of the composite as demonstrated in Figure 6.2-18. The specimen is a semicircular

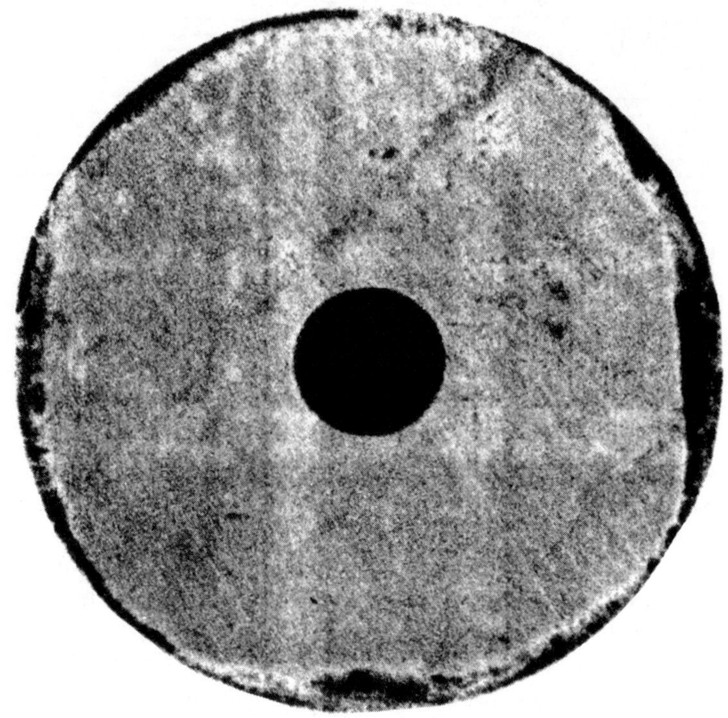

FIGURE 6.2-19. Circular disk with implanted disbond between plies.

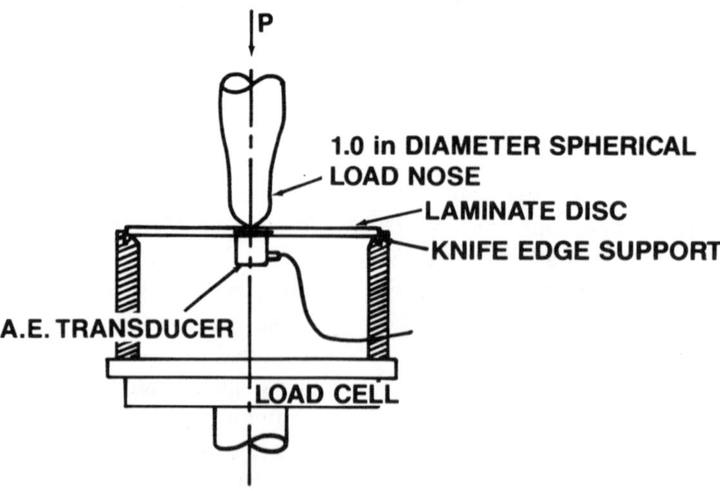

FIGURE 6.2-20. Static loading of circular disk.

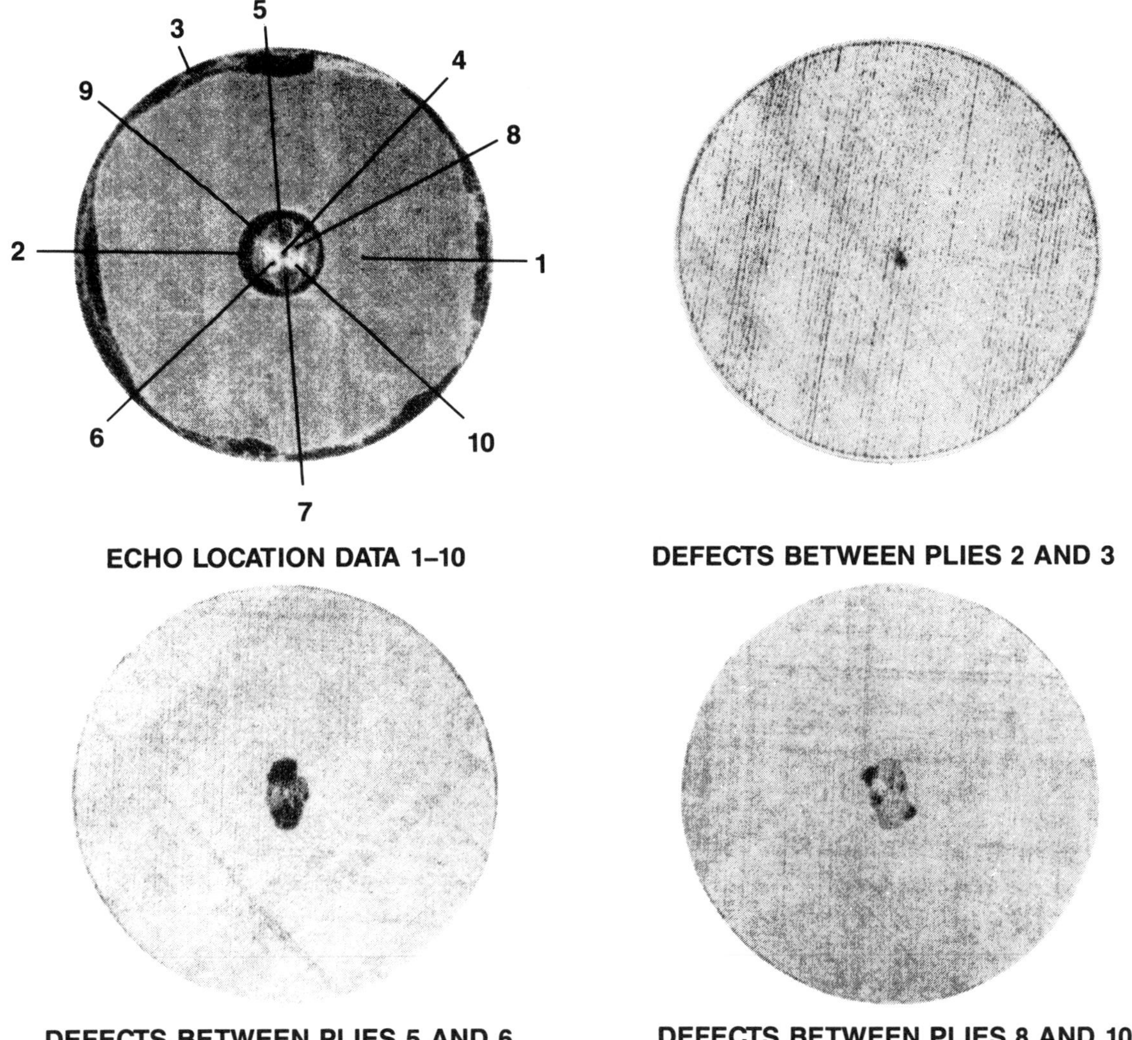

FIGURE 6.2-21. Lobe-shaped delamination isolation.

disk with a central disbond and a delaminated circumference. These features are clearly visible in the figure and suggest a method for large increment, high speed scanning.

6.2.4 Delamination Characterization

The ultrasonic pulse echo C-scan technique may be used to characterize interlaminar flaws resulting from impact, fatigue, or static loading. As an example, consider the ultrasonic study of a disk containing induced interlaminar damage.

For the study, a circular disk was fabricated with a circular disbond implanted at mid-plane in a $[0_2/\pm45_2/90_2]_{2s}$ graphite/epoxy laminate (see Figure 6.2-19). Damage was induced and propagated by loading the disk, supported on a circular knife edge in the center with a 25.4 mm (1 in) diameter spherical load base (Figure 6.2-20). Acoustic emissions [1,2] were monitored in the test section to detect propagation of the damage. The specimen was removed for C-scan inspection subsequent to a significant increase in acoustic activity. Figure 6.2-19 shows a C-scan of the disk prior to loading. The dark regions show the implanted flaw and

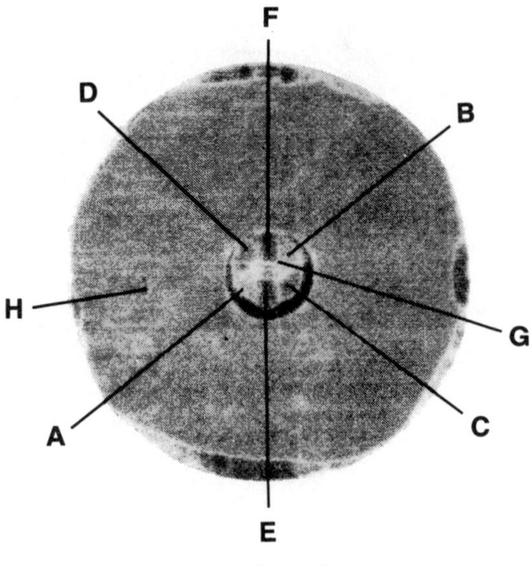

ECHO LOCATION DATA A–H

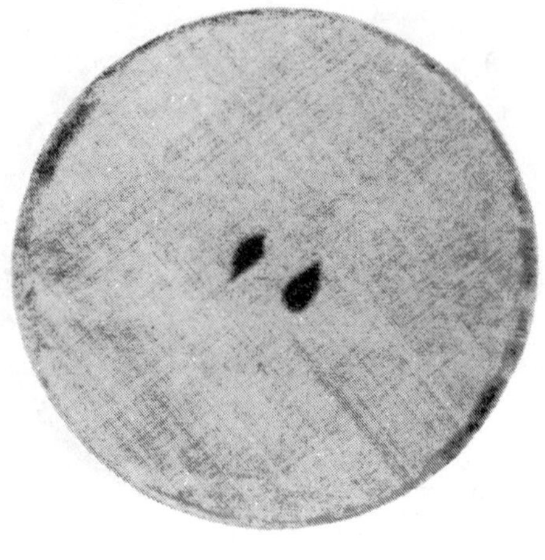

DEFECT BETWEEN PLIES 3 AND 4

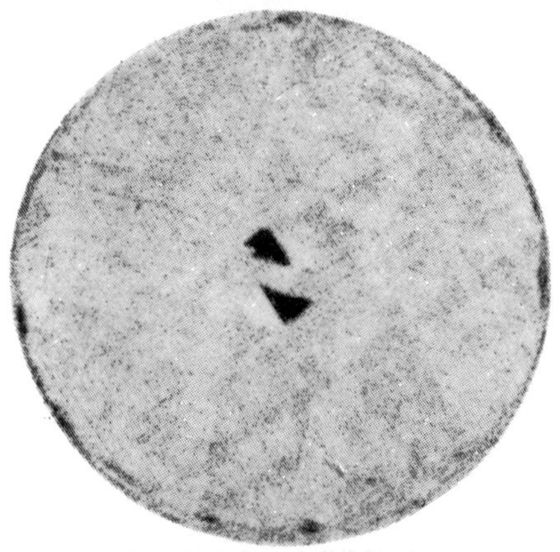

DEFECT BETWEEN PLIES 4 AND 5

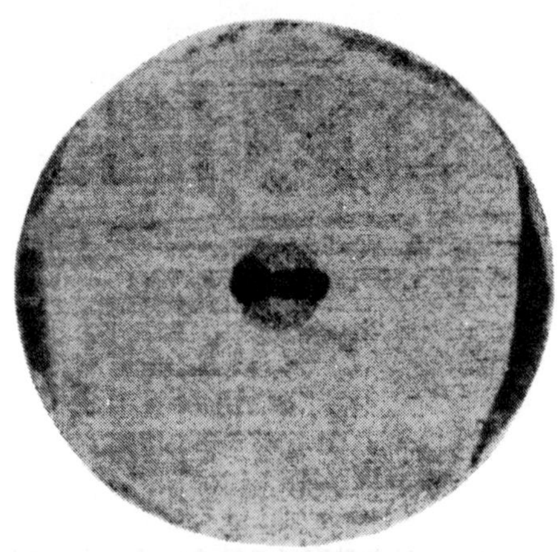

DEFECT BETWEEN PLIES 6 AND 7

FIGURE 6.2-22. Strip and wedge delamination isolation.

Table 6.2-1. Echo location data.

Sample Point	Interface Delay (μ sec)	Sample Point	Interface Delay (μ sec)
1	2.8	A	0.4
2	1.4	B	0.4
3	1.6	C	0.3
4	1.4	D	0.3
5	0.5	E	0.6
6	0.3	F	0.6
7	0.5	G	0.1
8	0.3	H	1.2
9	0.2	–	–
10	0.2	–	–

the presence of delamination damage around the edge that was caused by machining operations.

The C-scans in Figure 6.2-21 indicate that the delaminations in the top half of the specimen appear lobe-shaped and propagate in the direction of the fiber orientation below the delamination. Figure 6.2-22 shows strip- and wedge-shaped delaminations propagating below the implanted Teflon® disbond. Later studies have shown that the implanted disbond is not necessary to achieve the lobe and strip delaminations. The echo location data, summarized in Table 6.2-1, are used to determine the ply interface location of the delaminations. This method of location for damage is based upon time-of-flight data. In the subsection on Fourier analysis, a frequency-based method for delamination localization will be presented which offers improvements.

The photomicrograph of a sectioned disk, given in Figure 6.2-23, reveals delaminations that interconnect between the plies to form a staircase geometry. The interconnections are difficult to detect and display with ordinary C-scan techniques; however, the presence of these cracks can be detected using techniques described in a later subsection on digital image processing. Backscatter techniques have also been used on thin fatigue damaged materials, and the same type of through-the-ply crack is observed. The implanted Teflon® delamination is located at the center of the photo.

When the disk specimens were loaded until a second acoustic emission increase was observed, the ultrasonic C-scans in Figure 6.2-24 showed damage extending beyond the implant, although the lobe and strip character of the delaminations was maintained. Figure 6.2-25 summarizes these results. The delamination stacking sequence shown indicates lobe-shaped delaminations above the implanted defect and strip-shaped delaminations below. The delaminations interconnect between the plies through matrix cracks to form a staircase geometry. Impact induced delaminations differ from the above damage in that a damage cone is produced that shows little damage at the impacted surface and increases radially through the thickness [3].

Through this example, the power of some of the previously discussed techniques in ultrasonic NDE has been illustrated. So far, only analog methods have been employed. In subsequent subsections the increased performance of the method achieved through employing digital techniques will be discussed.

6.2.5 Transducer Considerations

The ultrasonic transducer plays an important role in determining the resolution of the ultrasonic inspection technique.

The four primary parameters associated with ultrasonic transducers are frequency, element diameter, focal length, and focal point diameter. These parameters interact with power and resolution such that as

1. Element diameter increases—frequency decreases
2. Frequency decreases—resolution decreases

TOP SURFACE—SPHERICAL LOAD INTRODUCTION

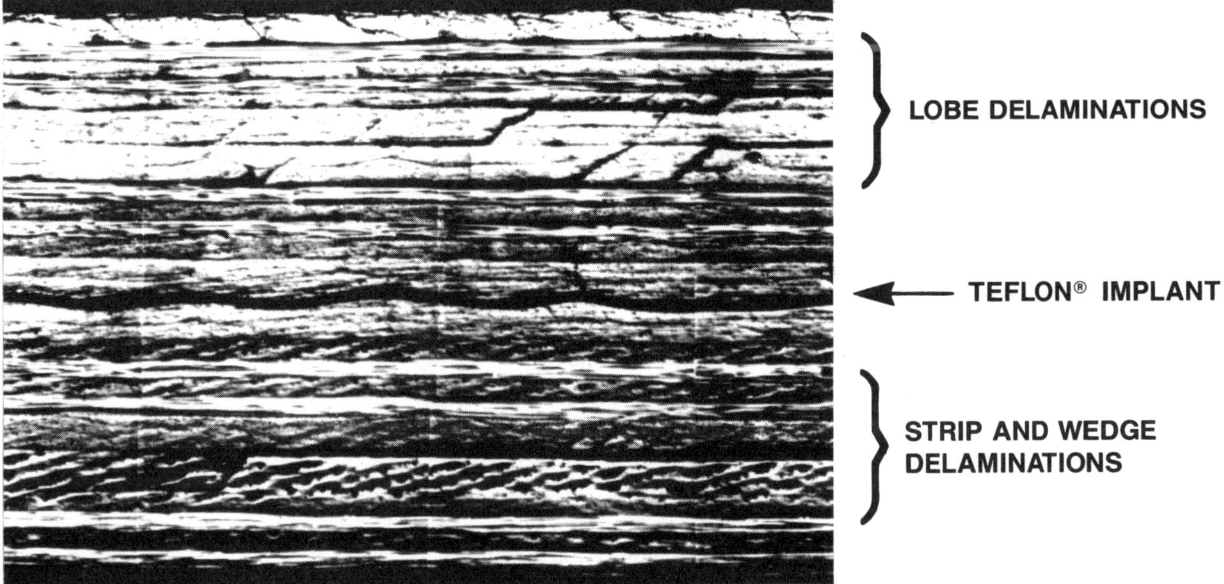

BOTTOM SURFACE—KNIFE EDGE SUPPORT

FIGURE 6.2-23. Photomicrograph of a sectioned disk subsequent to delamination propagation.

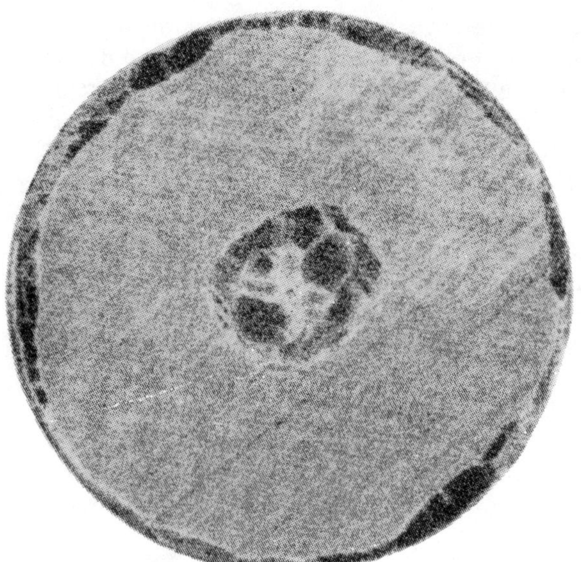

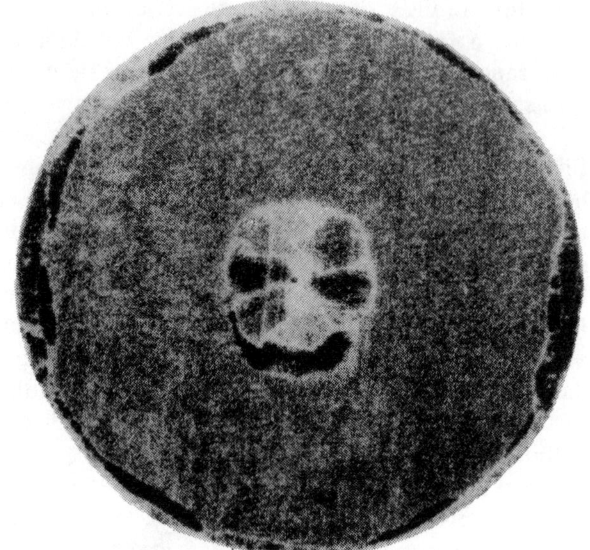

FIGURE 6.2-24. Multiple loading of disk samples.

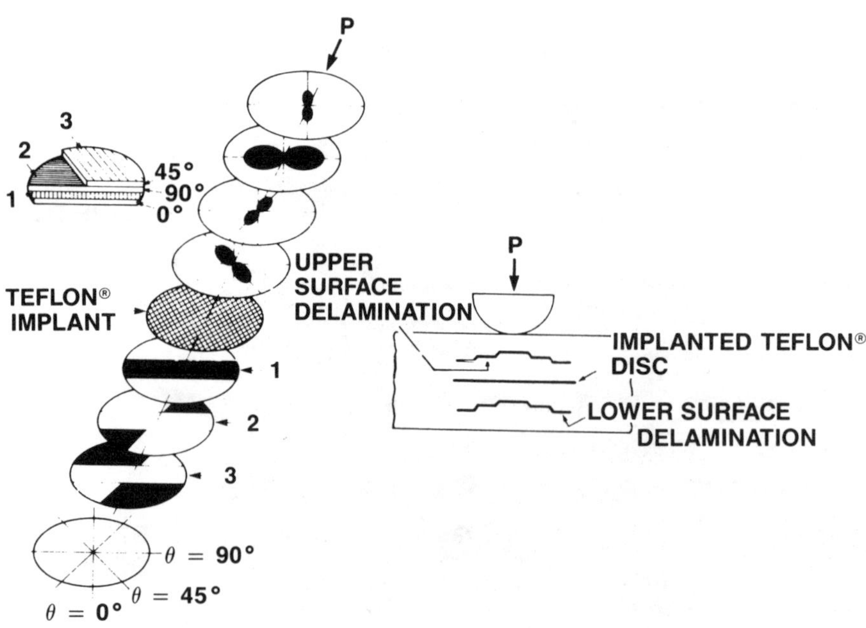

FIGURE 6.2-25. Delamination characterization summary.

3. Element diameter decreases—power decreases
4. Focal length increases—focal point diameter increases
5. Focal point diameter increases—resolution decreases.

These effects are illustrated in Table 6.2-2 for transducers 6.4 to 38.1 mm (0.25 to 1.5 in) in diameter and with frequencies ranging from 2.25 to 25 MHz.

Focal length is material dependent as well as a function of lens configuration, as indicated in Figure 6.2-26. Transducer selection and focal length determination become increasingly important as the material thickness increases. Flaws in thick laminates may not appreciably affect the ultrasonic waveform if the flaw is located too distant from the focal plane of the transducer.

Graphic techniques may be used to determine an optimum focal length for a given material, which equalizes the response from voids at selected depths as shown in Figure 6.2-27. The optimization was performed using a 15 MHz, 19.1 mm (0.75 in) element transducer with approximately a 102 mm (4.0 in) focal length in water. The receiver attenuation was set such that the back surface response just saturated at the one volt level. This attenuation level assumed that the response to a void was less than that of a solid to liquid interface. Measurements were recorded using positive

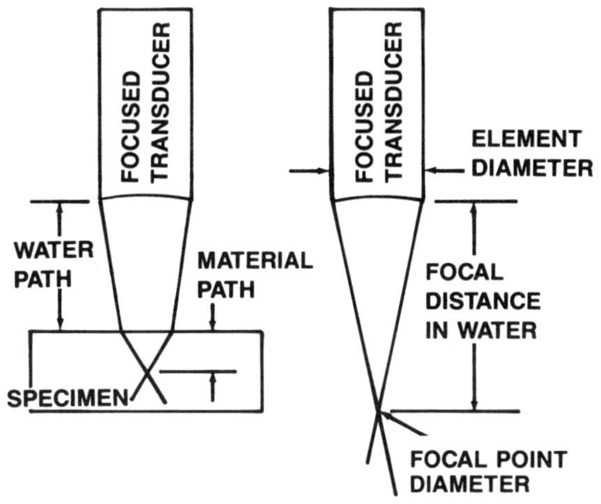

FIGURE 6.2-26. Focal length—material dependence.

peak data above 1.3 mm (0.05 in) diameter holes that were drilled on centers positioned at depths of 2.5 to 10.2 mm (0.1 to 0.4 in) at increments of 2.5 mm (0.1 in) normal to the edge of the composite. This hole depth implied that the actual defect depth to the top of the holes was offset toward the front surface of the specimen by a distance that corresponded to the radius of the hole. This effect could be seen clearly in the ultrasonic waveforms presented in the digital systems subsection. The method required that the amplitude response of each void at each depth be measured for a given transducer and material. It was then possible to select a focal length that equalized the response to voids at the 2.5 mm (0.1 in) and 10.2 mm (0.4 in) depths. The digital

*Table 6.2-2. Interaction of ultrasonic transducer parameters.**

Element Size (in)	Maximum Frequency MHz	Focal Length in Water (in)	Focal Spot Size Diameter (in)
1.5	2.25	4.2	0.13
1.5	2.25	6.5	0.23
1.5	2.25	9.2	0.37
0.75	15	2.5	0.05
0.75	15	4.0	0.09
0.75	15	5.2	0.15
0.50	20	1.7	0.04
0.50	20	2.7	0.06
0.50	20	3.7	0.12
0.375	20	1.3	0.03
0.37	20	2.2	0.05
0.375	20	3.6	0.08
0.25	25	0.8	0.02
0.25	25	1.5	0.05
0.25	25	2.2	0.07

*For Automation Industries immersion search units.

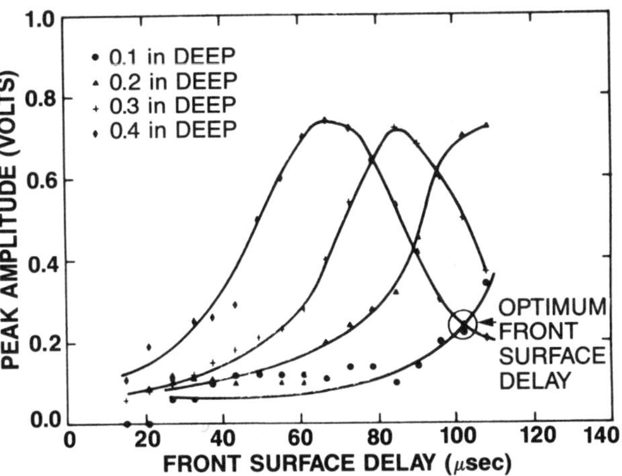

FIGURE 6.2-27. Focal length optimization.

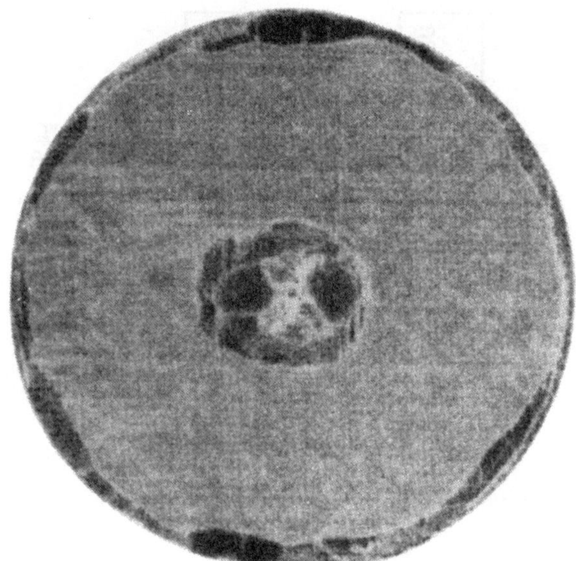

FOCUSED TRANSDUCER

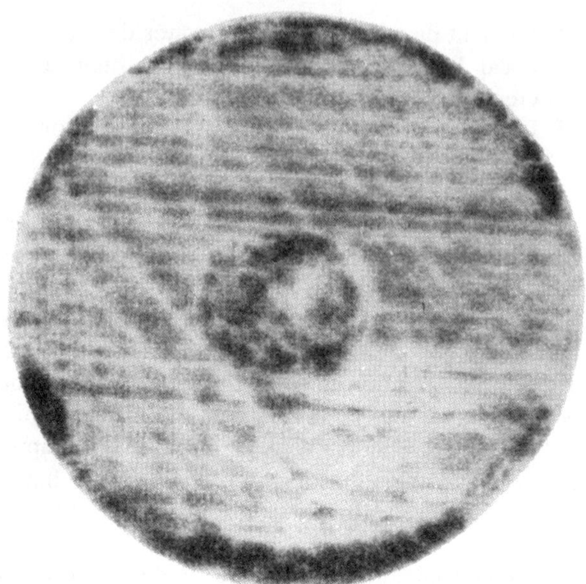

UNFOCUSED TRANSDUCER

FIGURE 6.2-28. Focused transducer C-scans.

systems approach discussed next will show that it is possible to equalize the response to voids at all depths using the optimum focal length determined from these measurements.

If unfocused transducers are used, the optimization technique does not apply, and void response will be much less dependent upon material thickness. Focused transducers produce sharp, well-defined images that allow for accurate mapping of delamination boundaries as demonstrated in Figure 6.2-28. Image clarity depends upon the focal parameters of the transducer; however, resolution depends upon the velocity of sound in the material and the frequency of the transducer as indicated in Figure 6.2-29.

In general, the highest frequency and shortest focal length possible should be used to form the C-scan. Material properties and structure may dictate the use of larger diameter, lower frequency transducers; material thickness may require longer focal lengths. The digital methods discussed next utilize the beneficial properties of focused transducers while minimizing the effect of focal distance.

5 MHz   LOW RESOLUTION

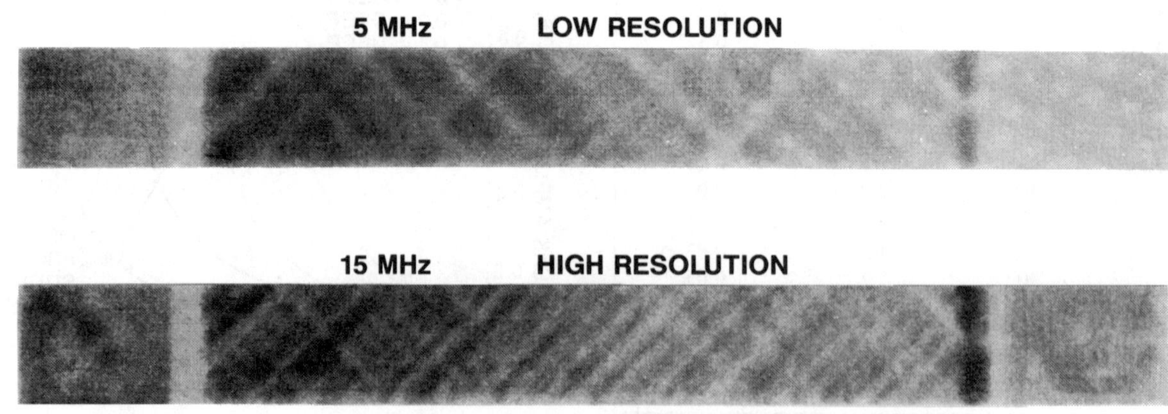

15 MHz   HIGH RESOLUTION

FIGURE 6.2-29. Frequency effects upon C-scans.

6.2.6 Digital Systems Approach

To fully utilize all the information available in the ultrasonic waveform and to equalize the response to voids at all depths in the composite, a more sophisticated form of ultrasonic analyzer was developed. This analyzer exists totally in Digital LSI 11 machine language code, except for a small gray scale control program that is written in 6502 machine code. The analyzer uses a Panametrics 5052 UA as a pulser/receiver and a Nicolet digital oscilloscope connected via a parallel interface to a Digital LSI 11 computer for waveform digitizing and analysis.

The Panametrics ultrasonic analog analyzer is capable of only one gated region and does not allow for gray scale conversions that would use low amplitude ultrasonic data. The gray scale is fixed to provide ten discrete gray levels in approximately 0.1 volt increments. This resolution implies that signals below 0.1 volt are ignored and that oscillations less than 0.1 volt are undetectable. Due to the subtle differences in the transition between adjacent levels of gray, it is doubtful whether oscillations less than 0.2 volts could be distinguished. The Panametrics analyzer specifies a maximum repetition rate of 5,000 interrogations per second; however, this rate reduces to 500 analog outputs per second because ten repetitions are required to drive the

analog output to the correct level. Furthermore, the analog signal appears as a sawtooth ramp that drops to zero volts after every ten repetitions. The repetition rate of the digital analyzer is a function of the number of gates and the gate widths, along with various delays associated with the types of information processing involved. If the digital analyzer is configured to perform the analogous functions of the analog analyzer, a repetition rate of 200 outputs per second can be realized. This assumes a digitizing rate of 50 Nsec per point and a gate width of 3.5 μsec. Unlike the sawtooth output of the analog analyzer, the output of the digital analyzer resembles a true analog signal that follows the peak output from the gate. Eight bits of data are transferred to the MCS 6502 microprocessor via a multiplexed six-bit bus; however, only six bits are used to form the analog output, which results in sixty-four possible gray levels.

The digital analyzer software allows for a maximum of fifty gated regions and provides a 250-word waveform buffer. Figure 6.2-30 shows a digitized ultrasonic waveform and identifies the digital analyzer parameters. The gated regions may be a minimum of 50 Nsec wide and may extend to the end of the waveform buffer. Gates may be specified in any region of the waveform and can be triggered to begin from a selected level. Each gated region is assigned a threshold value below which all information is ignored. The threshold levels extend

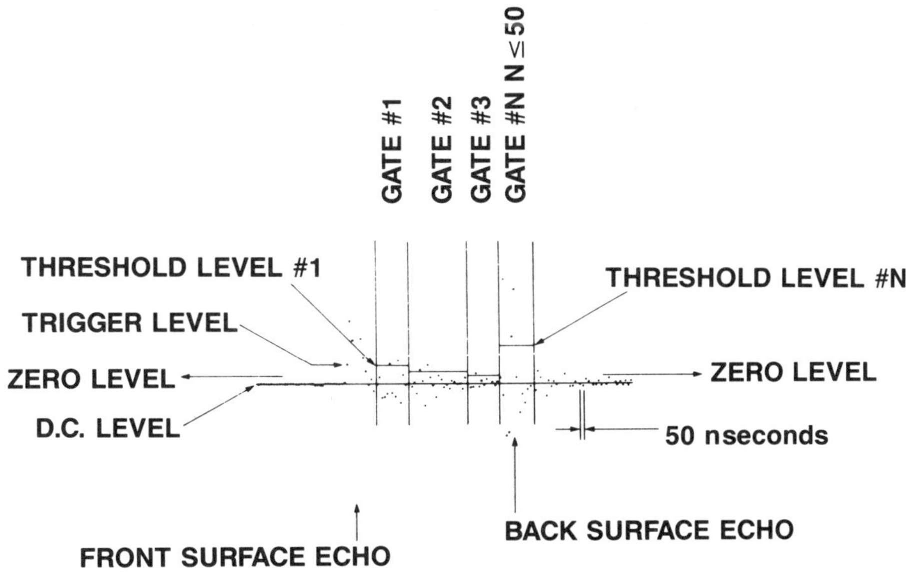

FIGURE 6.2-30. Digitized ultrasonic waveform identifying digital analysis parameters.

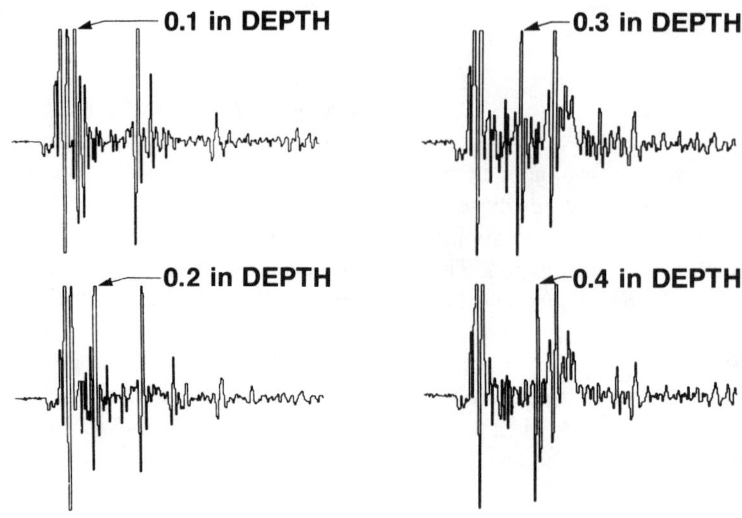

FIGURE 6.2-31. Void response equalization.

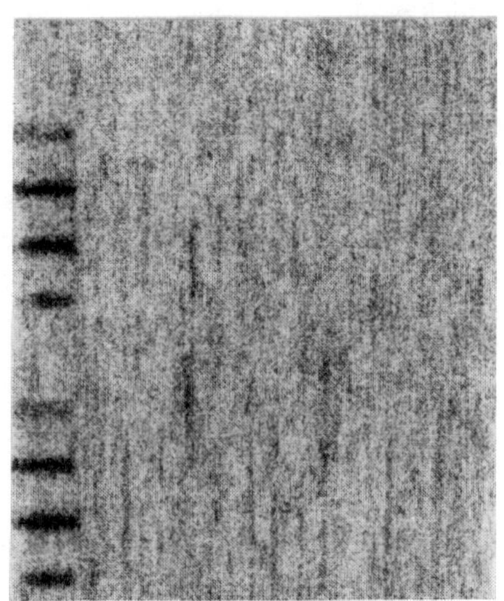

FIGURE 6.2-32. Analog and digital C-scans.

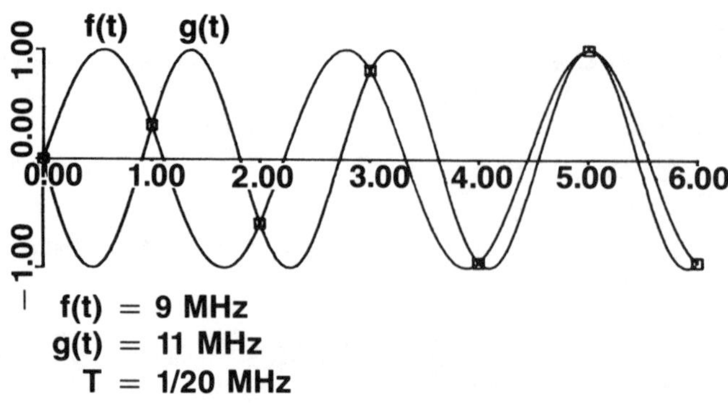

FIGURE 6.2-33. Aliasing in time sequential sampling.

equally above and below the zero level so that positive and negative data may be treated separately or together absolutely. The D.C. level may be used in conjunction with the threshold levels to obtain the desired gray levels and sensitivities; therefore, it is possible to set the amplifier gain and the appropriate levels in such a manner as to require the void responses to clip as shown in Figure 6.2-31. Thus, it is possible to equalize the responses to all voids regardless of depth and transducer focal length. The technique works best when the optimum focal length for the selected transducer is used; however, the range of usable focal length is greatly increased over that available with the analog analyzer.

The digital analyzer was set to have one gated region that corresponds to that of the analog analyzer, and ultrasonic C-scans were made over a portion of a specimen that contained holes drilled into the edge of a 12.7 mm (0.5 in) thick unidirectional composite, perpendicular to the fiber direction and at incremental depths. Figure 6.2-32 shows the digital and analog results. As expected, the responses to voids at all depths have been equalized and gray scale information is maintained. The responses to natural flaws appear to be enhanced in the digital image as well as having an improved edge definition.

A digital approach to ultrasonic nondestructive evaluation of composite materials appears to be the most versatile and cost-effective method for implementing a variety of inspection and measurement techniques. As requirements change with the advent of new composite materials, the digital system is adaptable through software development and program usage rather than through costly hardware additions. The digitizing of the ultrasonic waveform and the application of Fourier transform algorithms by general purpose computers provide analytical methods and spectral computations not possible by analog spectrum analyzers. The use of binary and color graphic displays eliminates many of the undesirable properties associated with conductive paper imaging. Reproducibility is increased by the elimination of variability in the display medium and by improvement of the signal-to-noise ratio made possible by digital image processing. The future application of color graphics to ultrasonic C-scan inspection and more sophisticated image processing techniques promise to provide a more quantitative basis for material evaluation and a facility for interpretation of ultrasonic composite material amplitude signatures.

Application of Fourier Analysis

The frequency spectrum of an ultrasonic waveform may be obtained by using an analog spectrum analyzer; however, an analog spectrum analyzer is a very specialized component in an ultrasonic system and is limited in the number of operations it can perform as well as in the form of data presented. A more versatile approach to forming the frequency spectrum of an ultrasonic waveform is by digitally sampling the waveform at a rate greater than twice the highest frequency component in the waveform and then evaluating the discrete Fourier transform of the sample points. The Fourier transform of a time dependent function $f(t)$ is $F(j\omega)$, defined in Equation (6.2-3):

$$F(j\omega) = \int_{-\infty}^{\infty} f(t) e^{-j\omega t} dt \qquad (6.2\text{-}3)$$

The discrete Fourier transform (DFT), $\overline{F}(j\omega)$, is defined as the zero order approximation to $F(j\omega)$:

$$\overline{F}(j\omega) = \sum_{n=-\infty}^{\infty} f(nT) e^{-j\omega nT} \qquad (6.2\text{-}4)$$

The zero order approximation in Equation (6.2-4) has been divided by T to give $\overline{F}(j\omega)$, since the differential dt becomes T in the approximation and t has been replaced by nT in the summation, where n is any integer and T is the sampling interval. Normally, $f(nT)$ is written as f_n, the nth sample point. $\overline{F}(j\omega)$, the discrete Fourier transform, is periodic with period $2\pi/T$. Since only a finite number of sample values is available, the DFT must be computed over the finite sum in Equation (6.2-5):

$$\overline{F}(j\omega) = \sum_{n=0}^{N-1} f_n e^{-j\omega nT} \qquad (6.2\text{-}5)$$

where N is the number of sample points. $\overline{F}(j\omega)$, which is periodic with period $2\pi/T$, possesses N complex elements, $N/2$ of which are independent. This function suggests computing $\overline{F}(j\omega)$ with a frequency spacing, ω:

$$\omega = 2\pi m/NT \text{ rad/sec}, \quad m = 0,1, \ldots, N-1$$
$$(6.2\text{-}6)$$

The equation for computing the DFT becomes

$$\overline{F}_m = \sum_{n=0}^{N-1} f_n e^{-j(2\pi nm/N)} \ , \ m = 0,1, \ldots , N-1$$

$$(6.2\text{-}7)$$

and the inverse DFT is defined as

$$f_n = \frac{1}{N} \sum_{m=0}^{N-1} \overline{F}_m e^{j(2\pi nm/N)} \ , \ n = 0,1, \ldots , N-1$$

$$(6.2\text{-}8)$$

Finally, the amplitude spectrum $|\overline{F}_m|$ may be obtained from the quadrature of the real and imaginary parts of $\overline{F}_m$:

$$\overline{F}_m = \sum_{n=0}^{N-1} f_n \cos 2\pi nm/N - j \sum_{n=0}^{N-1} f_n \sin 2\pi nm/N$$

$$(6.2\text{-}9)$$

$$\overline{F}_m = R_m + jI_m \qquad (6.2\text{-}10)$$

$$|\overline{F}_m| = [R_m^2 + I_m^2]^{1/2} \qquad (6.2\text{-}11)$$

The phase angle may be evaluated as

$$\phi = \tan^{-1}[I_m/R_m] \qquad (6.2\text{-}12)$$

The waveform must be sampled at a rate no less than twice the highest frequency component in the waveform, or aliasing will result because, given a sampling interval of T seconds, the sample values at frequency v and frequency $v + nT$ are samples of a waveform with the same frequency v. Figure 6.2-33 shows the ambiguity that results if the waveform is sampled at less than twice the highest frequency. The waveform $f(t)$ was sampled at a rate greater than twice its highest frequency, but $g(t)$ was sampled at less than twice its highest frequency. As shown in the figure, it is possible for the two waveforms to possess the same sample set for a given T. The criterion of sampling at a rate greater than twice the highest frequency is called the *Nyquist criterion*. The presence of frequency components above the Nyquist frequency or sampling slower than the Nyquist criterion allows has an effect on the DFT as shown in Figure 6.2-34(c). The figure shows how aliased and shifted spectra are added to the true spectrum to form an aliased version of the DFT. The true $F(j\omega)$ and $\overline{F}(j\omega)$ are shown in Figures 6.2-34(a) and 6.2-34(b) respectively. To avoid aliasing, the signal should be fed through a low pass filter with a sharp cutoff frequency and then sampled at twice the cutoff frequency. The examples contained in this report have been sampled at 20 MHz after first passing the signals through a fourth order Butterworth low pass filter with a 3 dB point of 10 MHz and a drop of 65 dB at 11 MHz. The computation time can be reduced substantially by the use of a fast Fourier transform algorithm, FFT. The FFT reduces the computation time by making use of the redundancy in the DFT. The DFT in Equation (6.2-7) requires $(N)^2$ complex products to transform N sample values into N frequency values. If we define a term W_N as

$$W_N = e^{-j(2\pi/N)} \qquad (6.2\text{-}13)$$

Equation (6.2-2) reduces to

$$\overline{F}_m = \sum_{n=0}^{N-1} W_N^{mn} f_n \qquad (6.2\text{-}14)$$

We may now separate Equation (6.2-14) into its even and odd components:

$$\overline{F}_m = \sum_{n \text{ even}} f_n W_N^{mn} + \sum_{n \text{ odd}} f_n W_N^{mn} \qquad (6.2\text{-}15)$$

Substituting $n = 2\ell$ into the first sum and $n = 2\ell + 1$ into the second sum, Equation (6.2-15) becomes

$$\overline{F}_m = \sum_{\ell=0}^{N/2-1} f_{2\ell} W_N^{2m\ell} + \sum_{\ell=0}^{N/2-1} f_{(2\ell+1)} W_N^{(2\ell+1)m} \qquad (6.2\text{-}16)$$

If we rewrite W_N^2 as follows,

$$W_N^2 = e^{-j2\pi 2/N} = e^{-j2\pi/N/2} = W_{N/2} \qquad (6.2\text{-}17)$$

and substitute into Equation (6.2-16), we obtain

$$\overline{F}_m = \sum_{\ell=0}^{N/2-1} f_{2\ell} W_{N/2}^{\ell m} + W_N^m \sum_{\ell=0}^{N/2-1} f_{(2\ell+1)} W_{N/2}^{\ell m} \qquad (6.2\text{-}18)$$

Each sum in Equation (6.2-18) is now a DFT with $N/2$ points, implying that each sum requires $(N/2)^2$ complex multiplications. There are also N complex multiplications required to multiply the W_N^m terms times the second sum. There are now $N + N^2/2 = N(1 + N/2)$ products compared with N^2 products in the original DFT. We now make the restriction that $N = 2^n$, where n is any integer. The simplest case of a 2-point FFT is shown in Figure 6.2-35(a). The structure shown in the figure is called a *butterfly*, and it is the smallest structure into which an FFT can be reduced. The numbers appearing in the butterfly are the powers of the W_N^m terms. Figure 6.2-35(b) shows this scheme extended to an 8-point FFT. Note that the 8-point FFT can be decomposed into two 4-point FFTs. This operation results in the number of complex products being reduced from the original N^2 products to $(N/2) \log_2 N$ products. This result becomes very important for large n. For example, if $N = 1{,}024 = 2^{10}$, the number of complex products is reduced from $N^2 = 1{,}048{,}576$ to $(N/2) \log_2 N = 3{,}549$.

The ultrasonic waveform is available only for a short period of time, approximately four microseconds (depending on the speed of sound through the material and the material thickness), and the sampling device has a maximum sampling rate limit. Therefore, if the sampling rate were 20 MHz, there would be only twenty samples available per microsecond of waveform. If a DFT were evaluated for these twenty samples, only ten different frequency points in the frequency spectrum would be evaluated because of the symmetry of the DFT. Thus, to obtain a good representation of the frequency spectrum between these ten points, it is necessary to "pad" the sample sequence with enough zeros to allow a continuous curve to be plotted between the original ten points. The result of padding the sample set with a different number of zeros is shown in Figures 6.2-36(a) and 6.2-36(b). In Figure 6.2-36(a), a 9 MHz sine wave was sampled at a 200 MHz rate yielding 133 sample points in six cycles of the waveform. Three hundred seventy-nine (379) zeros were added to the 133 sample points to give $2^9 = 512$ points in the FFT calculation. The frequency spacing of the resulting 256 different points allows the spectrum to be plotted up to 200 MHz, which is half the sampling rate. The first 10 MHz of the total spectrum is plotted in Figure 6.2-36(a). There are only 25 points in this 10 MHz region because there are only 256 points for the entire 100 MHz span of the total spectrum. If more zeros are

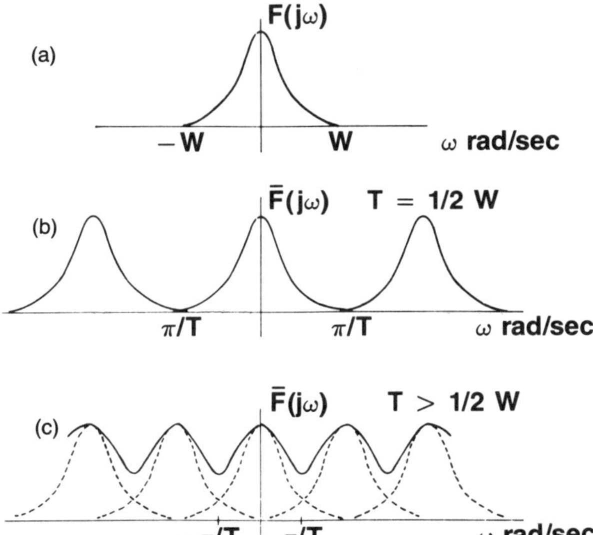

FIGURE 6.2-34. Aliasing in DFT.

added to the sample set to obtain $2^{12} = 4{,}096$ points in the FFT, the spectrum is evaluated at a much finer frequency spacing, as shown in Figure 6.2-36(b). There are now 204 points plotted in the spectrum instead of 25.

The sample device is digital, which implies that the sample values can take on only certain discrete values.

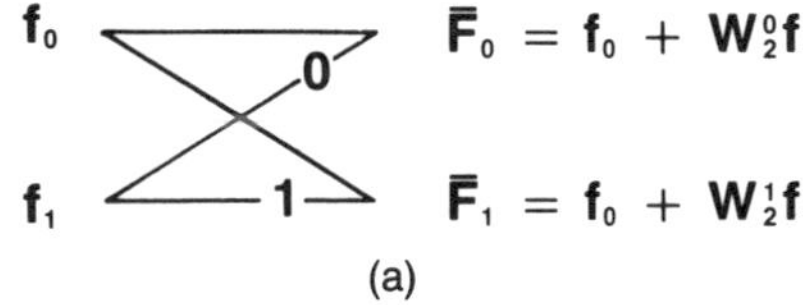

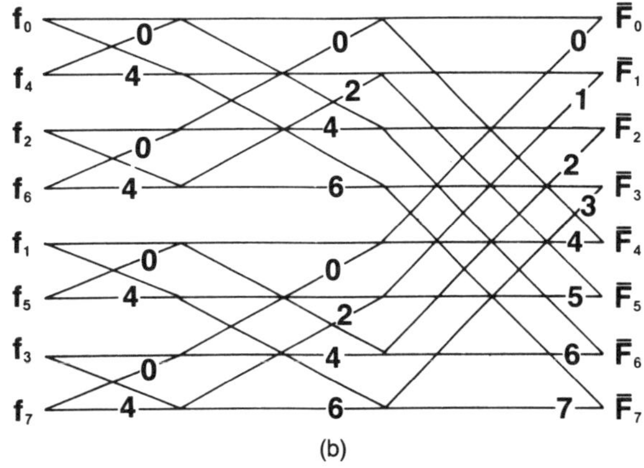

FIGURE 6.2-35. FFT representation: (a) 2-point; (b) 8-point.

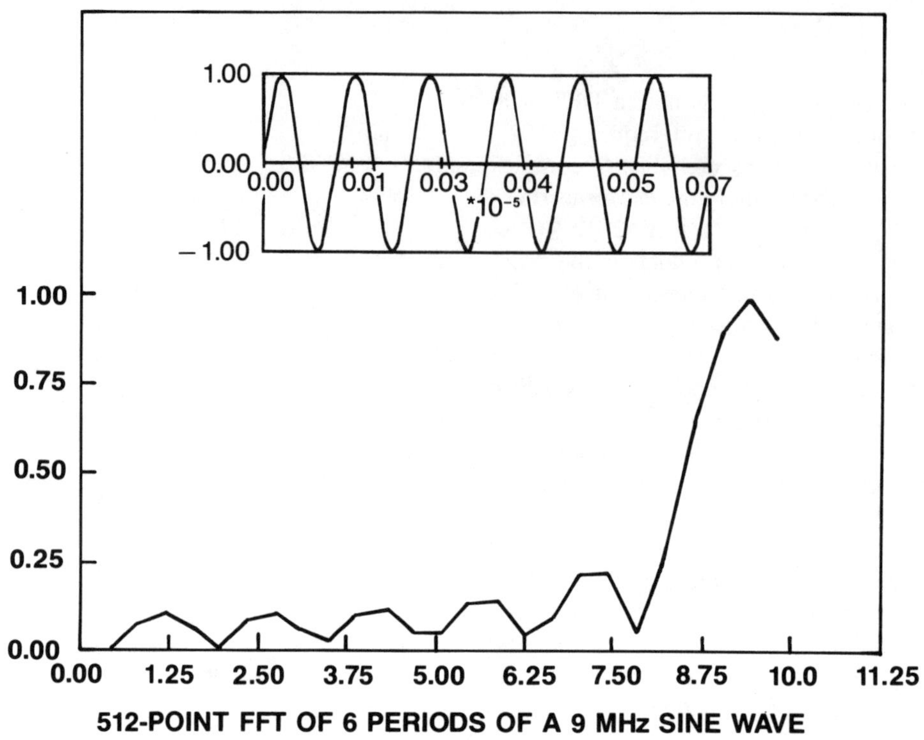

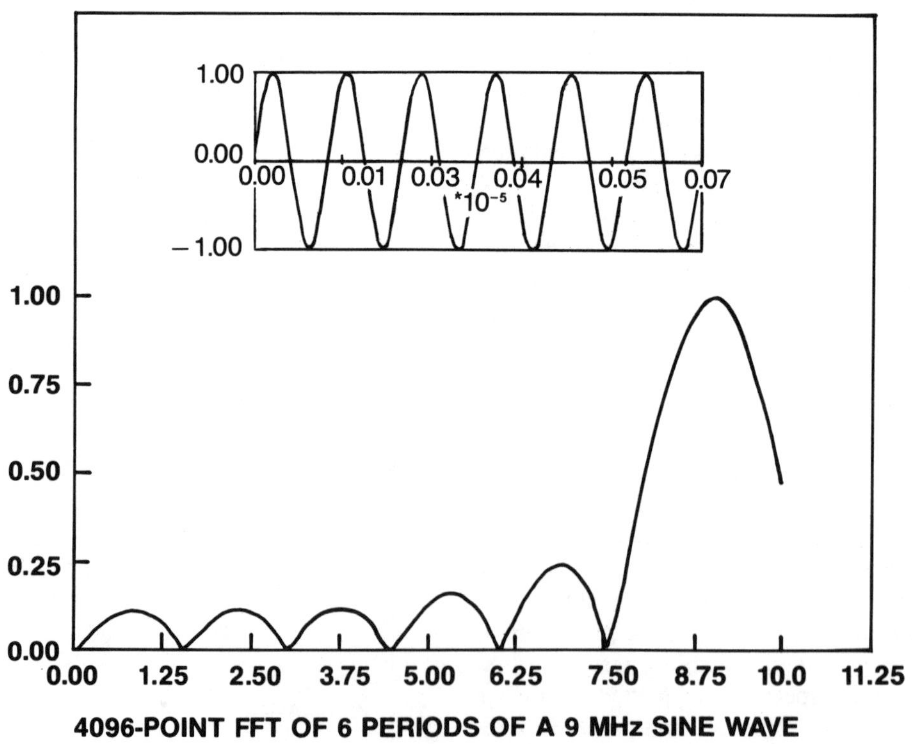

FIGURE 6.2-36. Interpolation of frequency spectrum.

The number of different values depends on the number of quantization levels. Figure 6.2-37 shows the effect of quantization on the FFT. The number appearing below each graph is the number of quantization levels between -1.00 and 1.00. Very slight differences in the resulting spectra are seen as the number of quantization levels decreases from 256 to 64; however, as the number of levels decreases from thirty-two to two, a definite increase in the D.C., zero-frequency component of the spectrum is seen. The side lobes of the FFT also become distorted as the number of quantization intervals decreases. It is interesting to note that, even with two quantization levels, the 7.5 MHz peak can be easily detected.

If the signal can be sampled for a length of time that is long compared to the period of the signal, a spectrum with sharp peaks can be produced. The size of the side lobes in the FFT will also decrease in amplitude. These effects are shown in Figure 6.2-38, in which the number of periods of a 5 MHz sine wave is increased from two to four. As the waveform is sampled for a greater number of periods, the peak in the spectrum converges

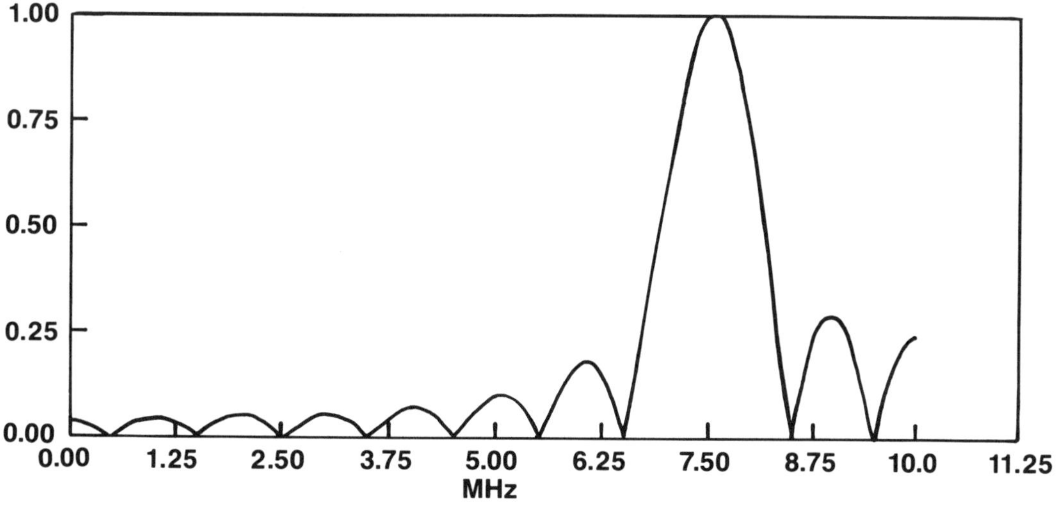

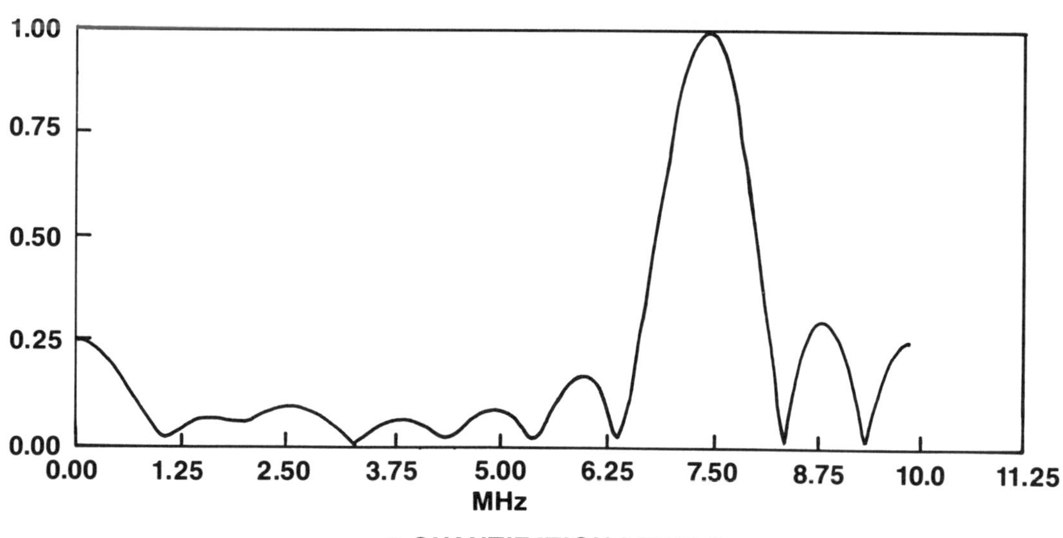

FIGURE 6.2-37. Quantization effects.

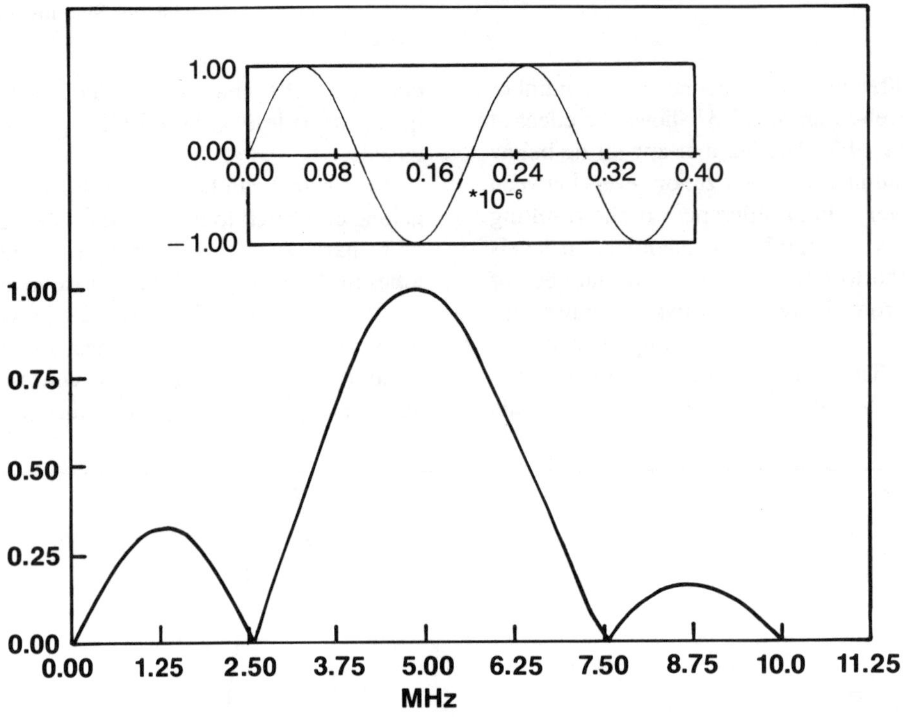

5 MHz WAVEFORM SAMPLED FOR TWO PERIODS

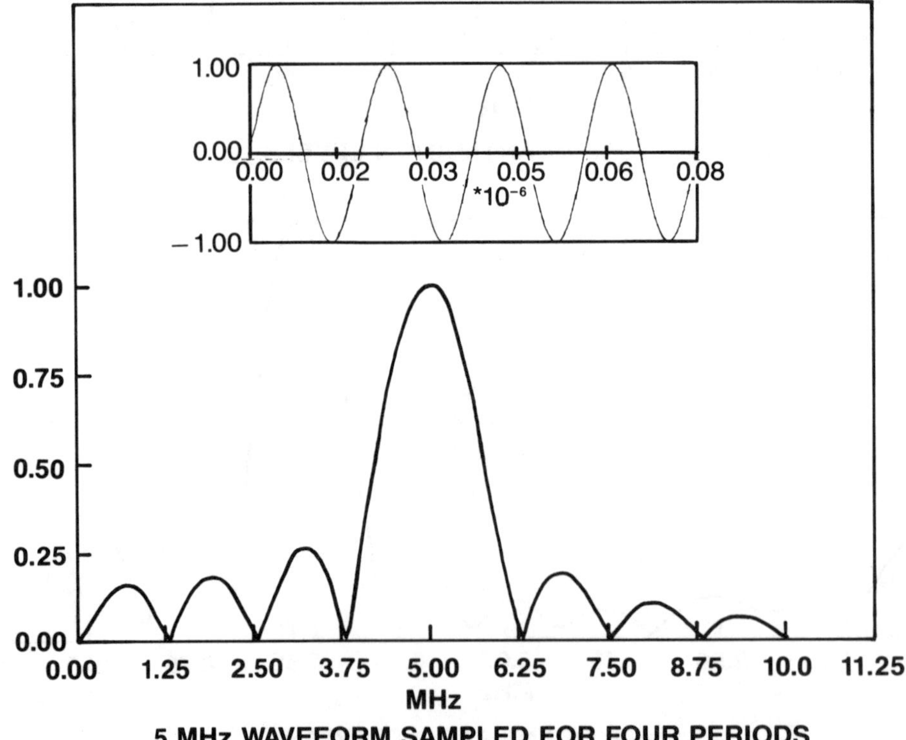

5 MHz WAVEFORM SAMPLED FOR FOUR PERIODS

FIGURE 6.2-38. Sampling duration effects.

and becomes sharp around the correct value. This effect is the same for both analog and digital computations and is due totally to finite signal duration. An application of frequency domain techniques is demonstrated in Figure 6.2-39. Zero padding was implemented in the frequency domain, and an interpolated version of the A-scan was produced upon inverse Fourier transforming the spectrum. The interpolated version of the waveform evaluated at 6.25 Nsec per point clearly shows the phase reversal produced by interface reflection.

The Fourier analysis of ultrasonic waveforms displays two characteristics. The shape of the curve in Figure 6.2-40 is due to the transducer's frequency response, and the oscillations are inversely proportional to the material thickness. The occurrence of a defect or discontinuity in the material will cause a corresponding change in the frequency spectrum. The presence of a defect at the center of the laminate results in the apparent thickness of the laminate to be half of the true thickness. This defect results in a frequency spacing,

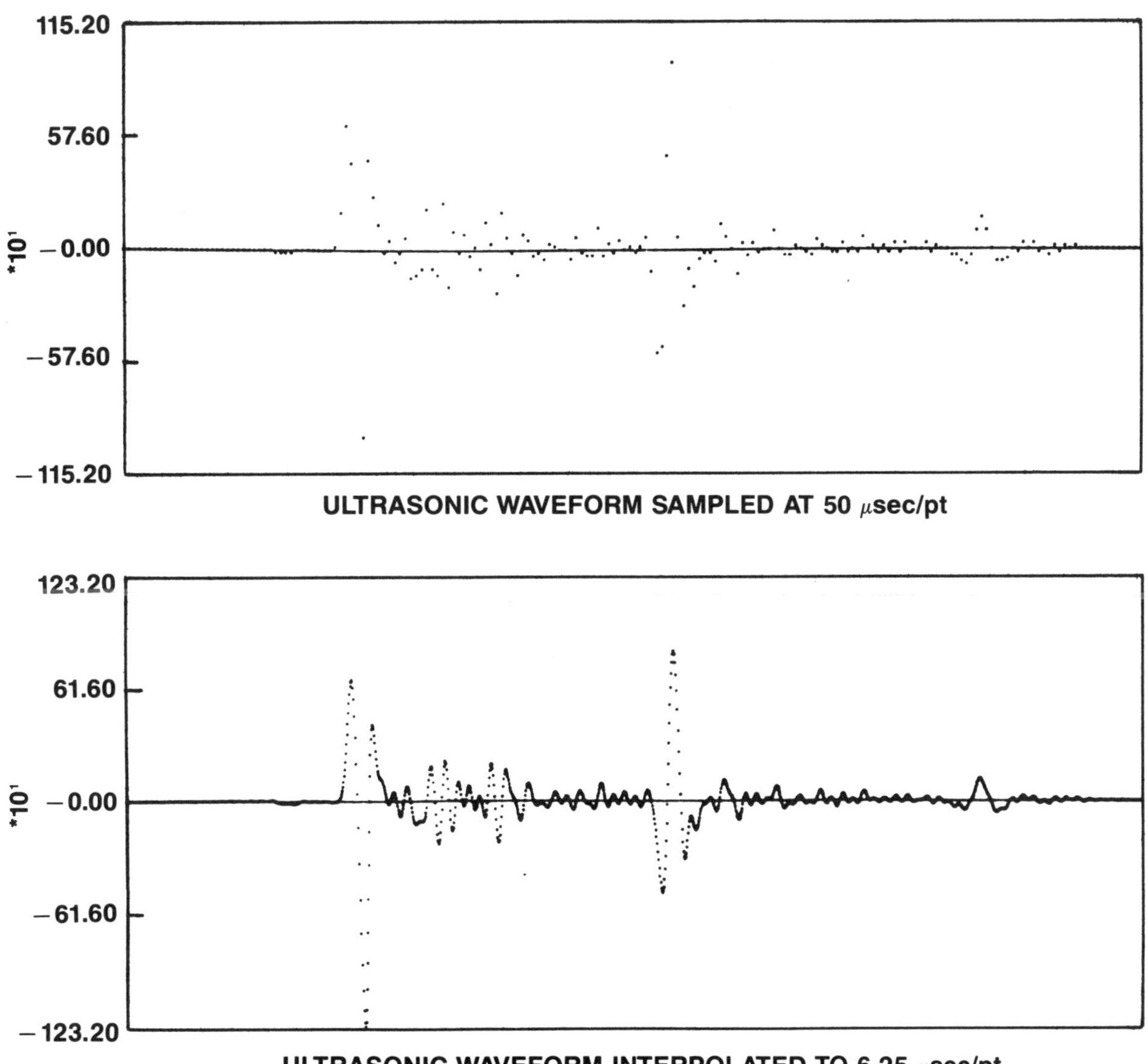

FIGURE 6.2-39. A-scan interpolation through Fourier analysis.

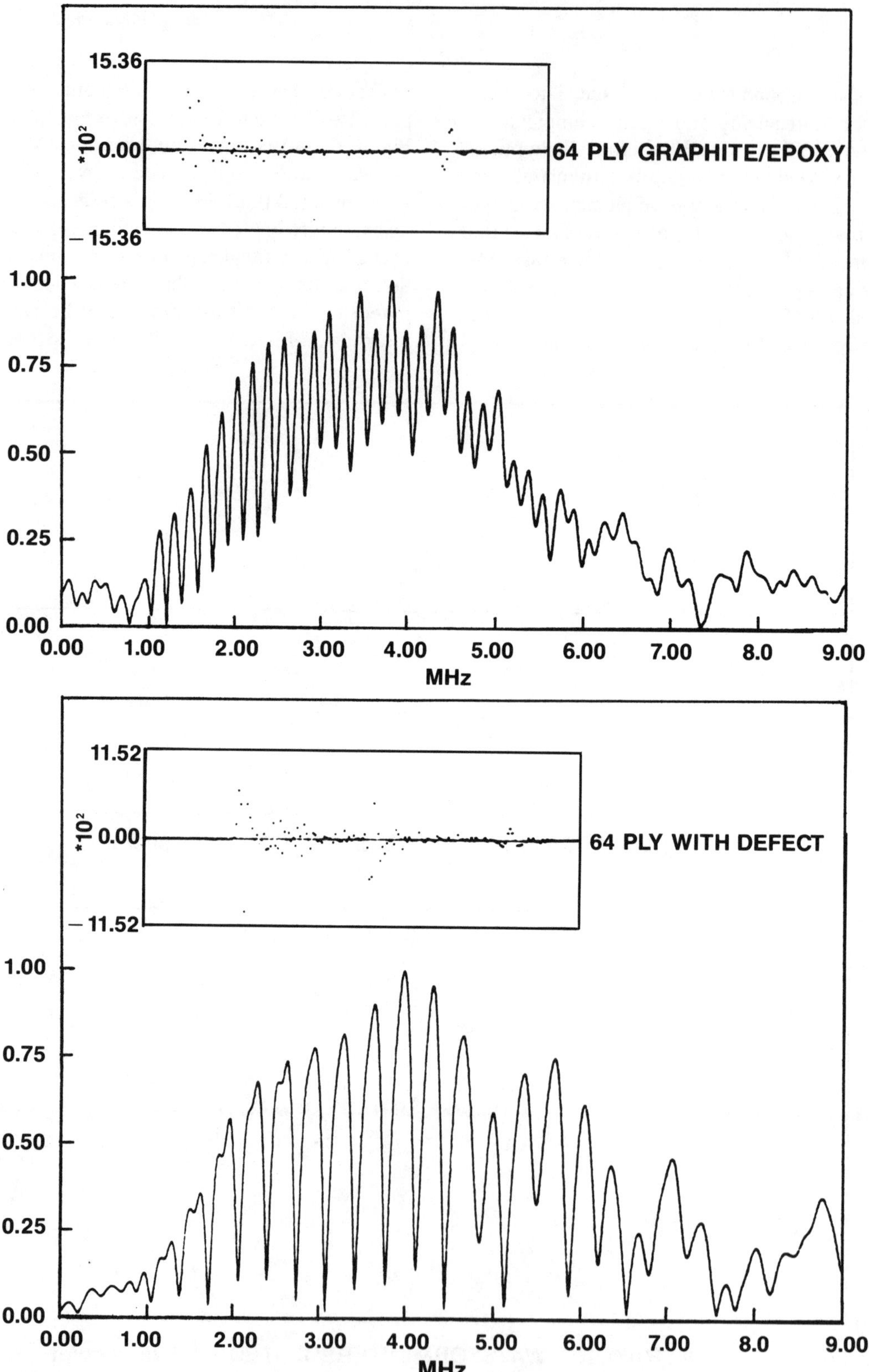

FIGURE 6.2-40. Fourier analysis of ultrasonic waveforms.

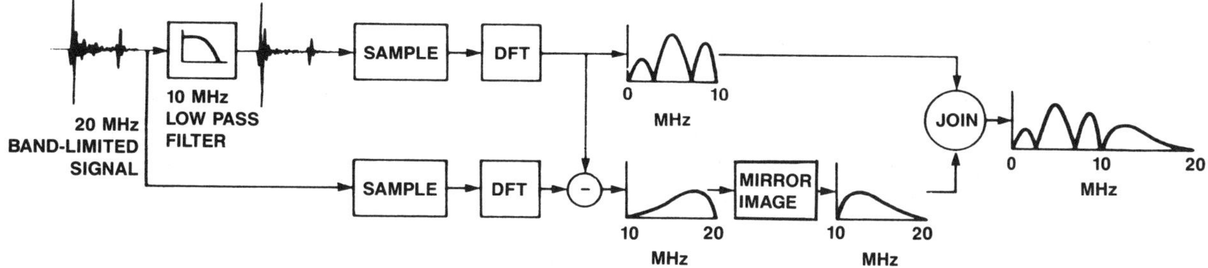

FIGURE 6.2-41. Selective filtering Fourier analysis at the Nyquist frequency.

Δf, of twice the value of Δf when a defect was not present, where

$$\Delta f = c/2d \qquad (6.2\text{-}19)$$

for velocity of sound c and thickness d.

The waveforms under investigation have frequency components up to 20 MHz, and (as mentioned previously) all signals are prefiltered such that they become band-limited to 10 MHz to prevent aliasing. A method devised to retrieve the frequency components up to 20 MHz without increasing the sampling rate is diagrammed in Figure 6.2-41. The spectral analysis is performed in two parts. First, the waveform is low pass filtered to 10 MHz, and the Fourier transform of the filtered and unfiltered signals is evaluated. The result-

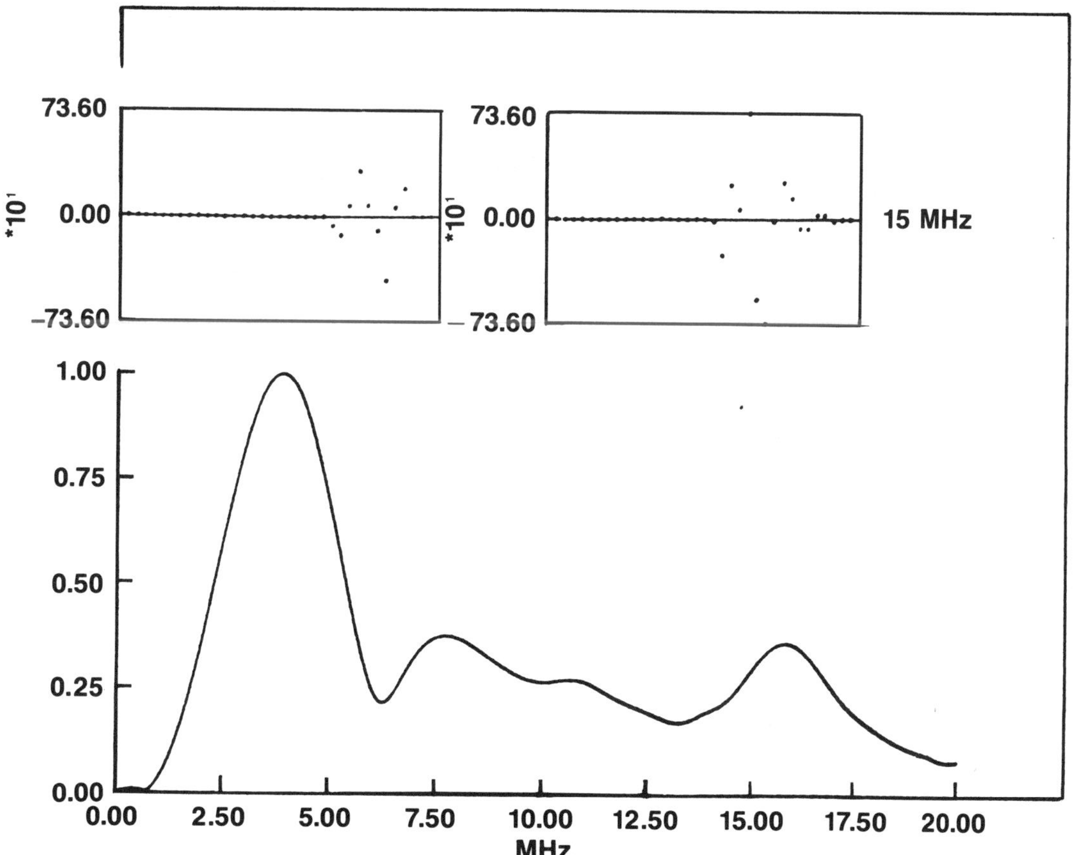

FIGURE 6.2-42. Transducer source spectrum evaluation.

ing spectra are subtracted, which produces one spectrum containing frequencies between 0 and 10 MHz and a second spectrum containing frequencies between 10 and 20 MHz. The mirror image of the second Fourier transform must be taken due to folding that occurs about the Nyquist frequency. This process may be continued to obtain spectra containing higher frequency components with the limiting factors of computation time and the frequency response of the sampling device. The source spectrum of a 15 MHz transducer was evaluated using this method, and the graphic output is shown in Figure 6.2-42. The results correspond well with those obtained from an analog spectrum analyzer in that there is a large, low frequency peak at 4 MHz and a lower amplitude peak near 15 MHz. The most significant result shown in Figure 6.2-42 is that the

slope of the curve is continuous at the 10 MHz folding frequency, which also tends to prove the validity of the algorithm.

Fourier analysis of ultrasonic waves in composites provides a mechanism for precise measurement of defect depth and allows for velocity-of-sound measurements to be made as a function of frequency. Figure 6.2-43 shows the result of applying the Fourier transform to the digitized ultrasonic waveform four times. The upper left graph in the figure shows the digitized waveform of a sixty-four-ply graphite/epoxy laminate with a disbond located in the center. The waveform was initially digitized at a 50 Nsec per point rate and Fourier transformed. The resulting frequency spectrum was then properly padded with a sufficient number of zero data points such that, when the inverse Fourier trans-

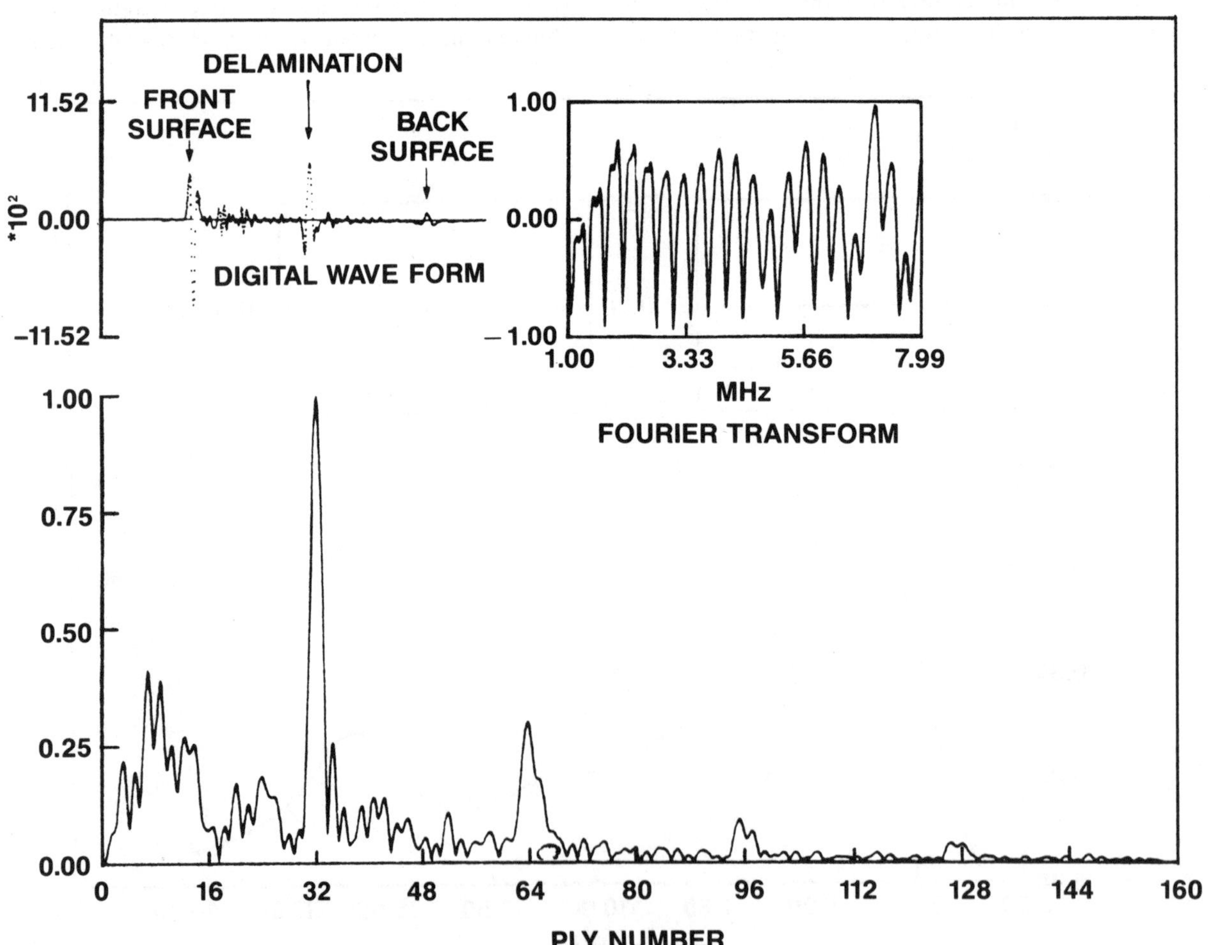

FIGURE 6.2-43. Echo peak enhancement.

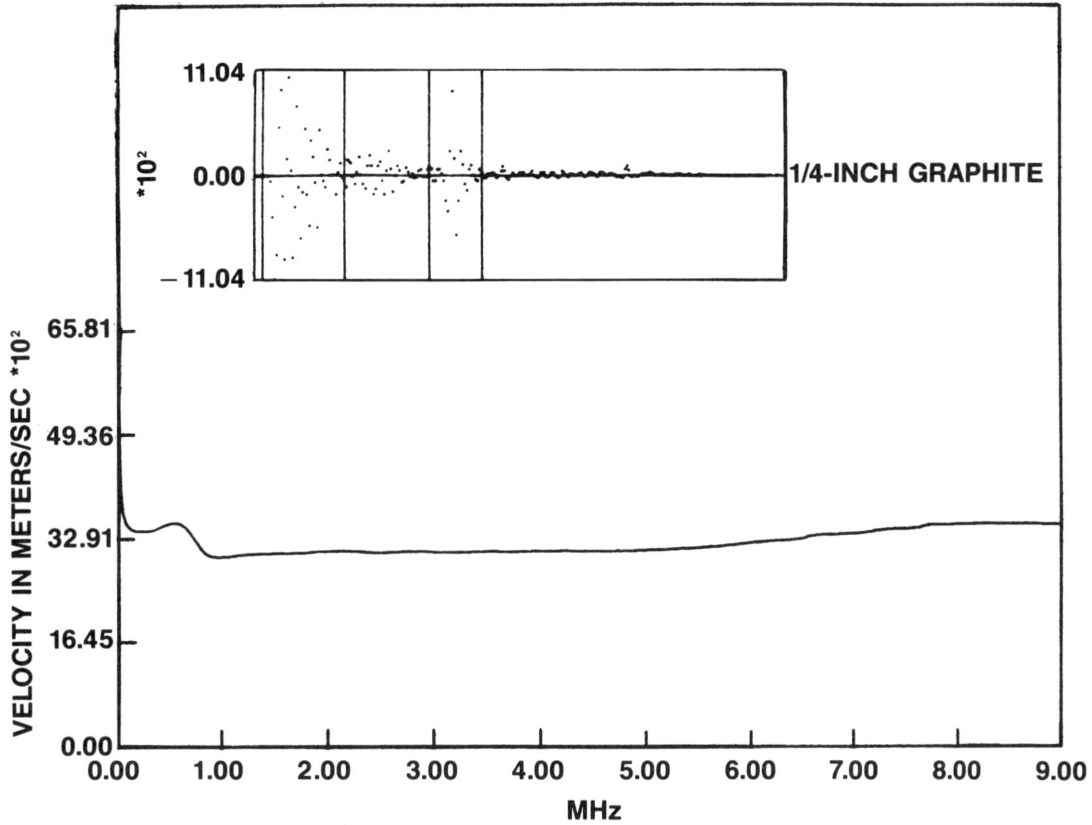

FIGURE 6.2-44. Phase velocity of sound measurement.

form was performed, the waveform was reconstructed and evaluated at 6.25 Nsec per point. A third Fourier transform was performed on the initial pulse received by the transducer. The transform of this pulse contains information related to the frequency response of the transducer alone. A deconvolution is performed by the complex division of the transform of the total ultrasonic waveform and that of the initial pulse. The trend and zero frequency term are then eliminated by a least square fit that is subtracted from the data. The resulting amplitude spectrum is shown in the upper right graph of Figure 6.2-43. The oscillations that appear are directly proportional to the velocity of sound in the material and are inversely proportional to material thickness. A final inverse transform is performed on the deconvolution and results in the time domain representation shown at the bottom of Figure 6.2-43. The sharp peak precisely locates the disbond at the ply thirty-two interface.

The group velocity of sound $c(\omega)$ may be evaluated as a function of frequency by computing the phase portion of the Fourier transform for the front, ϕ_0, and back, ϕ_1, surface reflections rigidly fixed in time. This phase information can then be used in Equation (6.2-20) along with the thickness of the composite, d, to form a velocity distribution such as that shown in Figure 6.2-44.

$$c(\omega) = \frac{2d\omega}{[\phi_1(\omega) - \phi_0(\omega)]} \qquad (6.2\text{-}20)$$

The waveform at the top of the figure was digitized at a dwell time of 50 Nsec per point, and the vertical lines indicate the region over which the Fourier transform was computed for each reflection. The fluctuation in the zero to 1 MHz range is indicative of group velocity dispersion.

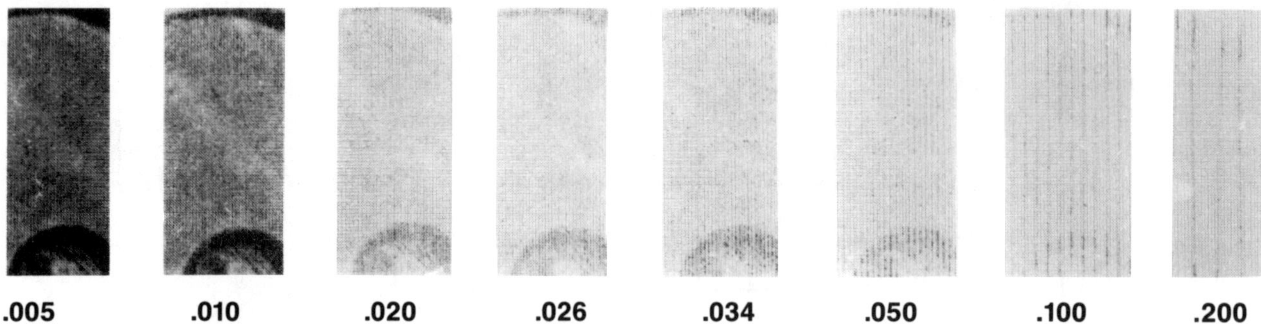

FIGURE 6.2-45. Image quality degradation as a function of scan line spacing.

Digital Image Processing

Most ultrasonic inspection facilities incorporate a mechanical linkage between the scanner and the display device. The display medium is normally an electrically conductive paper, and the image is burned into the surface layers.

There are several factors that affect image quality, speed, and versatility of mechanical systems. To decrease the time it takes to scan a specimen, either the velocity of the transducer or the spacing between scan lines may be increased. However, increasing the velocity of the transducer causes an undesirable decrease in image density or gray level. Increasing the scanner line spacing causes the image quality to degrade very rapidly as the display line spacing exceeds the display spot size or line width, even if the line spacing is adequate for the image detail that is of interest. The image degradation that results is shown in Figure 6.2-45. This degradation is a consequence of the mechanical linkage between the scanner and display. The linkage also restricts magnification between the scanner and display to 1:1. Dry paper systems that do not have a mechanical linkage between the display and the scanner permit image magnifications and reductions without the degradations shown in Figure 6.2-45. However, image magnification requires a longer scan time than 1:1 images.

The use of conductive paper is further complicated

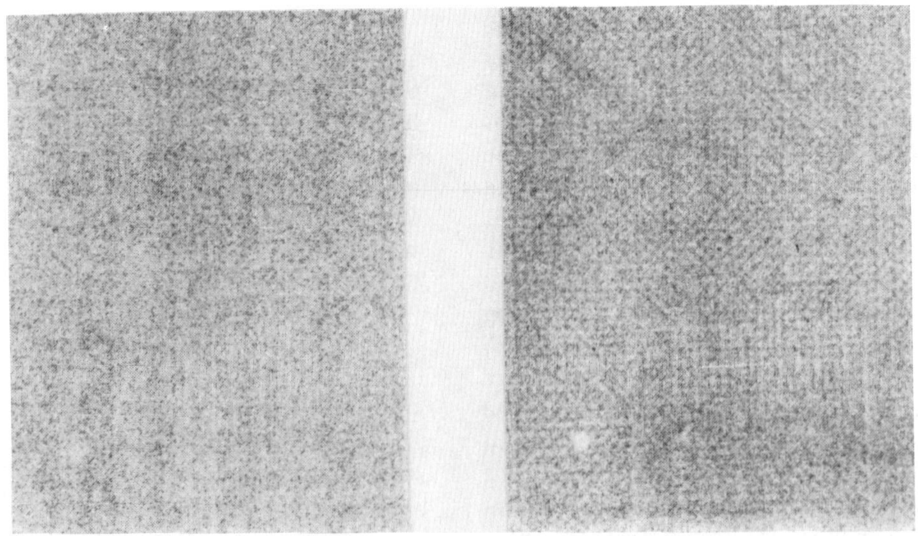

FIGURE 6.2-46. Variability in image reproducibility due to variation in the recording medium.

due to the variability in the paper that results in contrast changes as well as changes in the overall gray scale. Figure 6.2-46 shows ultrasonic C-scans of a woven composite material. The two scans were made with the same control settings, and the paper was from the same manufacturer but from two different lots. This variability in the paper makes reproducibility very difficult. In addition, the paper is very susceptible to scratches and smudges during storage.

A digital system with a binary display eliminates the problem associated with conductive paper imaging and also increases system flexibility. Image magnification is possible since there is no longer a mechanical linkage between the scanner and the display. Additional detail not visible in the 1:1 image is presented by allowing samples taken at a higher density than those required for the 1:1 image to occupy the full display area. This image differs from enlargement of the data in any sub-region of memory to occupy the full display area. Image quality may be kept constant in the digital system regardless of the transducer velocity on the increment between successive passes across the material. Therefore, decreased scan times are possible. Since the display parameters may be set independently of the scan parameters, the resolution of the C-scan is limited only by the parameters of the sound field. The digital system also allows for image storage on magnetic tape. Electronic ordered dither techniques can be employed to represent gray levels on a binary display device. A binary display is attractive for reasons of low cost, low memory requirements, and simplified hard copy.

Representing image information in digital form immediately makes image processing possible. Gray scale modification and contrast enhancement are simple to implement and may eliminate the need to rescan materials to obtain satisfactory results. Image enhancement techniques can facilitate feature extraction. In particular, edge enhancement should greatly increase the ability of the viewer to interpret the results of the ultrasonic inspection.

This discussion describes a digital system that utilizes ordered dither techniques to present and store ultrasonic C-scans. The ordered dither technique is discussed in detail, and sample gray scales and digital C-scans are presented, which show the advantages of a digital system with a binary display. The system may be implemented at relatively low cost on any mechanical scanner with the addition of an axis encoder, micro-

processor, and display. Systems that are computer interfaced and have digital control over the transducer location need no modification and require only the addition of the microcomputer and display.

The test facility diagrammed in Figure 6.2-47 is based on a modified TekTran mechanical immersion scanner with a minimum scan line spacing of 0.025 mm (0.001 in). There is an optical encoder attached to the search tube assembly. The encoder monitors the Y-axis position along the scan line and delivers a pulse for every 0.025 mm (0.001 in) of travel. The encoder signals produce a relocatable origin on the Y-axis. When the search tube crosses this origin, the start of a new scan line is indicated. The origin signal and the position signal are sent to tristate buffers before they reach the microprocessor system.

The ultrasonic analyzer produces a pulsed signal that has an amplitude corresponding to the amplitude of the largest echo within a region of the total ultrasonic waveform selected by gating. This signal is then low pass filtered to produce an analog signal. A five-bit quantization is performed using active elements that do not require interface or handshake signals. The five-bit signal is then tristate buffered before it reaches the microprocessor. The ultrasonic transducer used in this work is an Automation Industries SIL 20 model 57A8118. This is a 20 MHz transducer with a 43.2 mm (1.7 in) focal length in water and a 1.02 mm (0.04 in) diameter focal point. The ultrasonic element is lithium sulphate, 12.7 mm (0.5 in) in diameter.

The system uses an MOS Technology MCS 6500 microprocessor to perform all the data sampling and display operations. The microprocessor performs all the necessary tasks and stores the entire image in 4,096 eight-bit words of memory. The microprocessor is attached to a Tektronix graphics terminal for control and display of the ordered dither images. An impact dot matrix printer could be substituted for the terminal as could any binary display device.

Electronic ordered dither is a technique by which images composed of various gray levels may be represented on a binary display device. The technique requires dividing the display area into a grid of elements that each contain a 4×4 array of binary display cells as shown in Figure 6.2-48. Each cell within an element is assigned a threshold value, and this matrix of threshold values is the dither signal. In the normal ordered dither image construction, a sample value is taken for

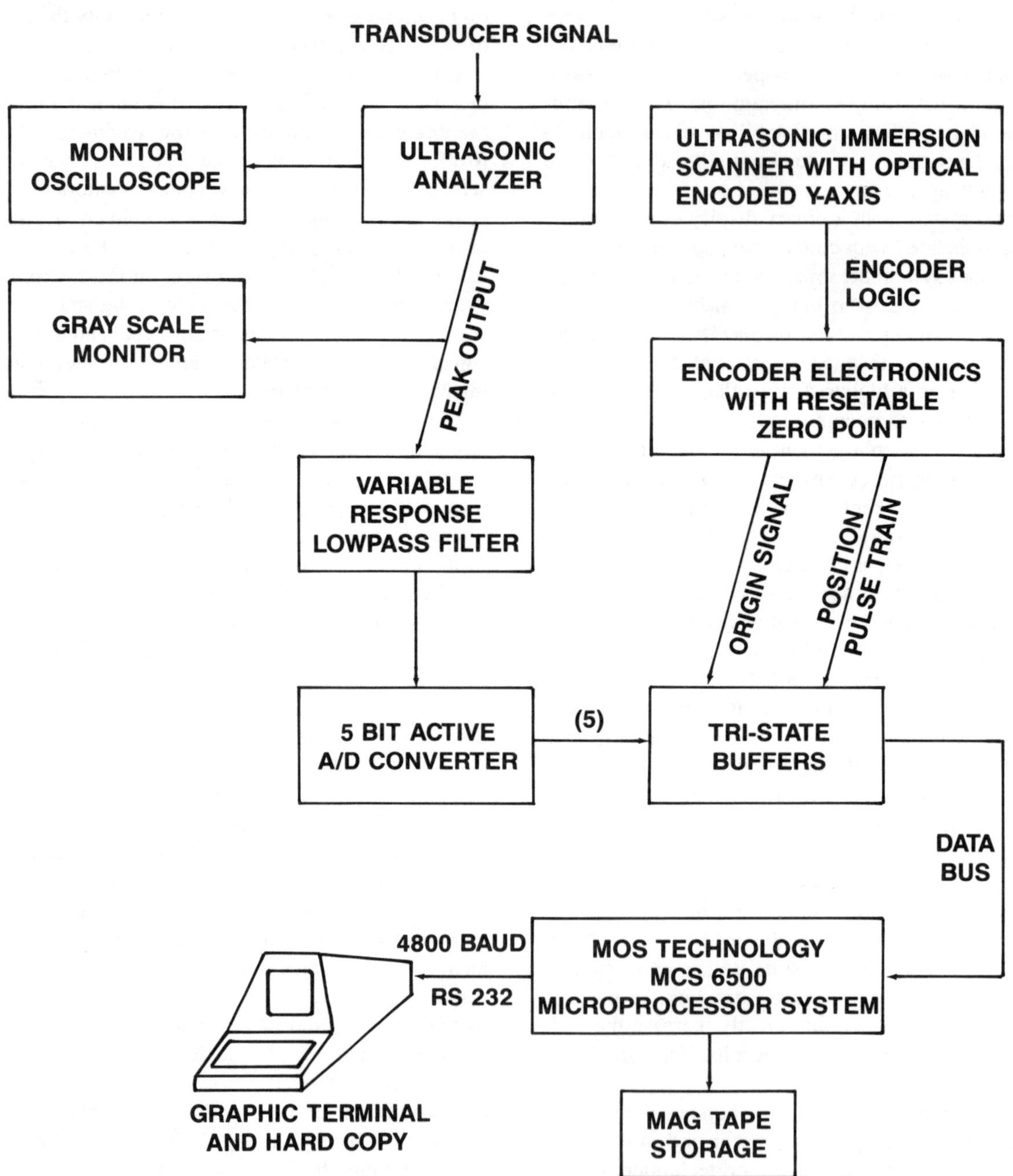

FIGURE 6.2-47. Data sampling and display system.

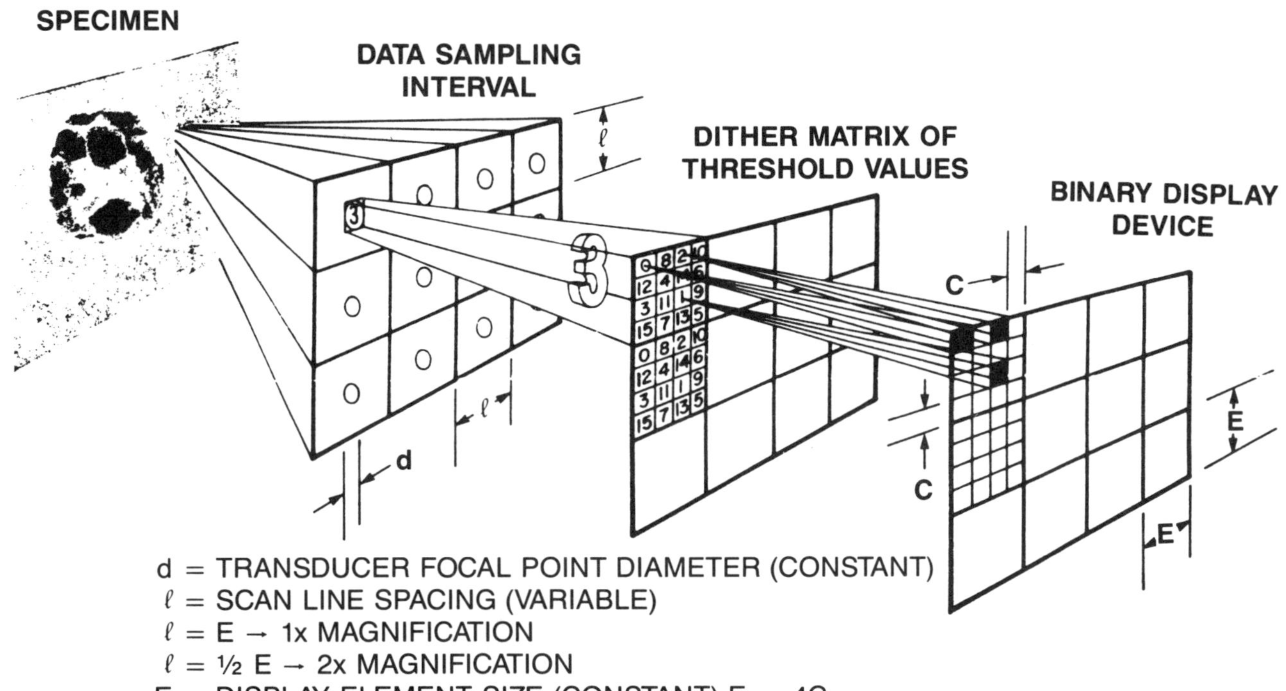

d = TRANSDUCER FOCAL POINT DIAMETER (CONSTANT)
ℓ = SCAN LINE SPACING (VARIABLE)
ℓ = E → 1x MAGNIFICATION
ℓ = ½ E → 2x MAGNIFICATION
E ≡ DISPLAY ELEMENT SIZE (CONSTANT) E = 4C
C ≡ SIZE OF ADDRESSABLE BINARY DISPLAY CELLS (CONSTANT)

FIGURE 6.2-48. Ordered dither image construction and magnification.

10	15	2	6
4	5	9	11
14	1	13	3
0	8	7	12

DITHER MATRIX A

0	8	2	10
12	4	14	6
3	1	11	9
15	7	13	5

DITHER MATRIX B

FIGURE 6.2-49. Dither matrices.

DOT PROFILE A

DOT PROFILE B

FIGURE 6.2-50. Dot profiles for two dither matrices.

FIGURE 6.2-51. C-scan of semicircular disk, 1×.

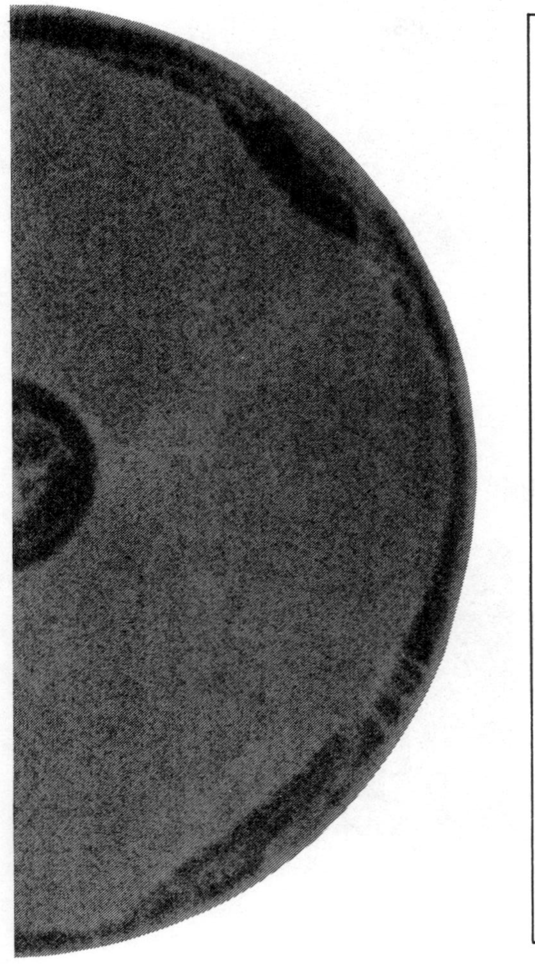

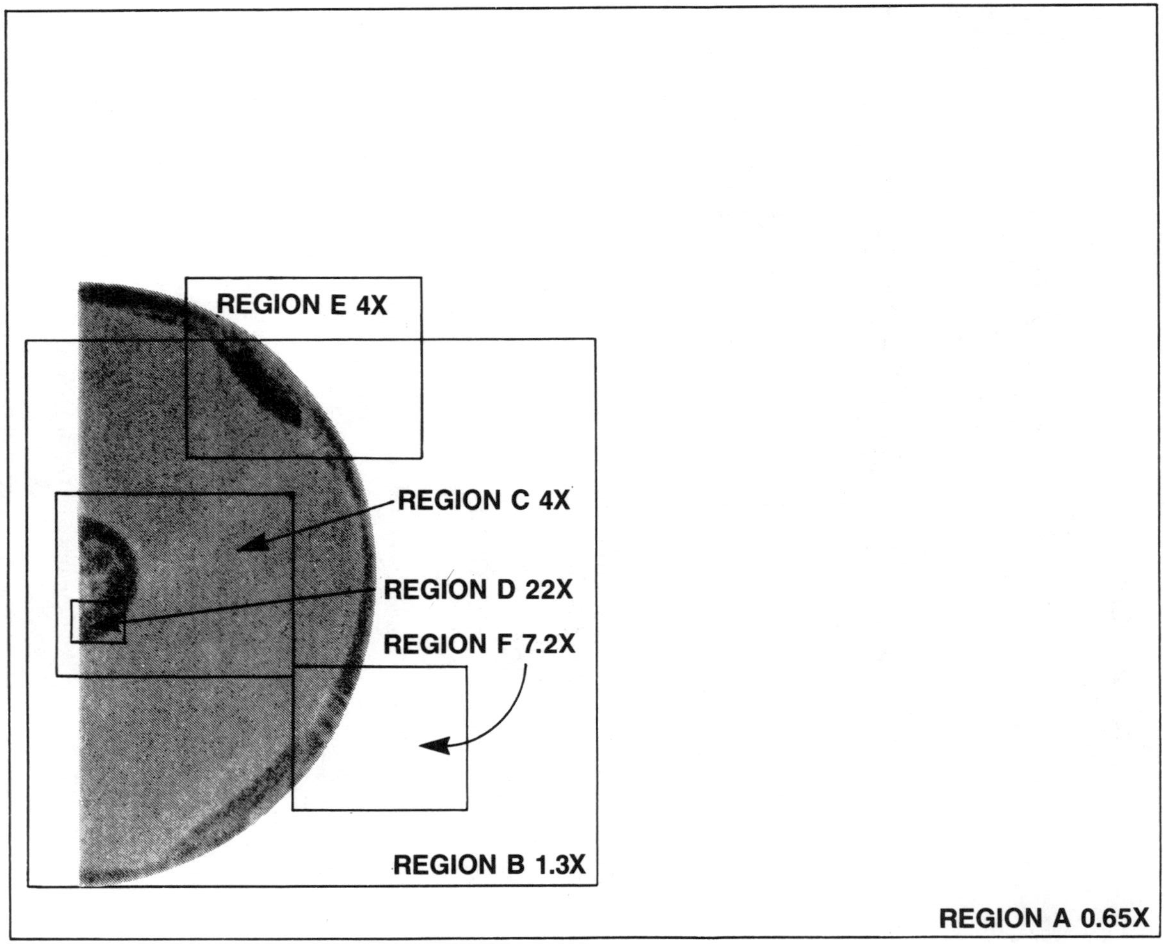

FIGURE 6.2-52. C-scan of regions of interest and magnification constants.

each of the cells within the matrix. If the sample value at a point within the matrix exceeds the threshold value of that point, the cell is illuminated; otherwise, it remains dark. This result is $2^{16} = 65,536$ possible dot patterns for a 4×4 matrix; it requires that sixteen sample values be taken. In order to take sixteen sample values per element, four scan lines per element must be made. Therefore, a factor of four savings in time can be achieved by making only one scan line per display element and taking one sample value per element. This single sample value is then compared to all sixteen threshold values within the matrix, as illustrated in Figure 6.2-48 for the element with sample value three. This handling of the data results in seventeen possible dot patterns that require five binary bits to store digitally. The primary effect upon image quality of taking only one sample per element is a loss in edge definition, since the displayed image resolution is equal to the display element size. This loss should be weighed against the factor of four time savings as well as the savings in the amount of memory required to store the image. Since this work is focused upon increasing the speed and versatility of an ultrasonic C-scan facility, the single sample per element method will be explored.

The dither matrices shown in Figure 6.2-49 will be used to represent the data. The ordering of the threshold values in dither matrix A was optimized using an inverse distance weighting, and dither matrix B was optimized using an inverse distance squared weighting. For each dither matrix, Figure 6.2-50 shows the family of binary patterns used to represent the levels of gray between white and black that will be referred to as the *dot profile*. Dot profile B contains many geometric shapes such as light and dark crosses and other patterns that are composed of lines of dots intersecting at right angles. These geometric patterns are not so visible in dot profile A. Also, the transition between gray levels is smoother in dot profile A than in dot profile B.

The specimen used in this study is half of a 127 mm (5.0 in) graphite/epoxy disk. A 25.4 mm (1.0 in) circular two-ply Teflon® delamination 0.051 mm (0.002 in) thick was implanted at the mid-plane of a $[0_2/\pm45_2/90_2]_{2s}$ laminate. The disk was statically loaded with a 25.4 mm (1.0 in) spherical load introduction and supported on a circular knife edge. The loading produced new delaminations above and below the implanted defect. During the machining process prior to testing, delaminations were produced around the circumference of the specimen. The use of a specimen with both service-initiated and implanted defects, as well as a geometry that includes both straight lines and curves, shows the capabilities and limitations of the electronic ordered dither ultrasonic imaging technique.

An ultrasonic C-scan was made of the semicircular disk. This C-scan, made with a line spacing of 0.254 mm (0.010 in), is shown in Figure 6.2-51. Figure 6.2-52 indicates the regions of interest and the corresponding magnification constants associated with each region. Figure 6.2-53(a)-(f) shows these six regions as displayed using ordered dither with the two dither matrices. The $4\times$ magnification of region C and the $22\times$ magnification of region D both indicate a vertical crack separating two delaminated regions in the lower third of the implanted semicircular delamination. This crack is visible on the initial 1:1 C-scan; however, it was not noticed until the magnified images were produced. The magnified images show the boundaries of the delaminations much more clearly than is possible in a standard 1:1 C-scan formed on a conductive paper. Delaminations that appear to be continuous in the 1:1 image show hairline cracks running across the delaminations in the magnified images. These cracks are visible in photomicrographs of the material and appear to interconnect delaminations on various interfaces with ply cracks oriented at 45° to the plane of the delaminations.

The major differences between dot profile A and dot profile B can be seen in the images of region C at $4\times$ magnification. Dot profile A gives a much smoother transition between gray levels, and edges of the delaminations are much better defined. Examples of the increased edge definition are also visible in the other figures, especially around the circular edge of the disk. Dot profile A appears to give the best overall image of the C-scan, especially when viewed from a distance. Dot profile B is particularly useful when looking for transitions in gray level. These transitions are clearly visible at close range. In general, both profiles have benefits in the interpretation of ultrasonic C-scan data. Magnification of the regions of interest on the specimen also greatly improves the ability of the viewer to interpret the results. The mechanical scanner permits the image to be magnified to slightly above $100\times$; however, the transducer does not have this high a resolution, so useful magnification is limited to approximately $40\times$. Ordered dither images that are formed using a single sample value per element are referred to as *pulse surface area modulation* (PSAM) images. PSAM images can be formed on a dot matrix printer as mentioned previously and provide superior image quality as shown in Figure 6.2-54.

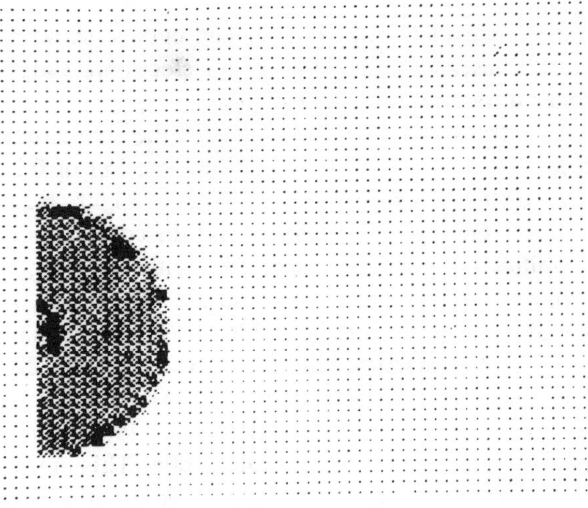

4 × 4 DOT PROFILE A

4 × 4 DOT PROFILE B

Region A
Magnification 0.65X
Scan Line Spacing 5.1 mm (0.20 in)

4 × 4 DOT PROFILE A

4 × 4 DOT PROFILE B

Region B
Magnification 1.3X
Scan Line Spacing 2.54 mm (0.10 in)

FIGURE 6.2-53. Ordered dither C-scans of selected regions.

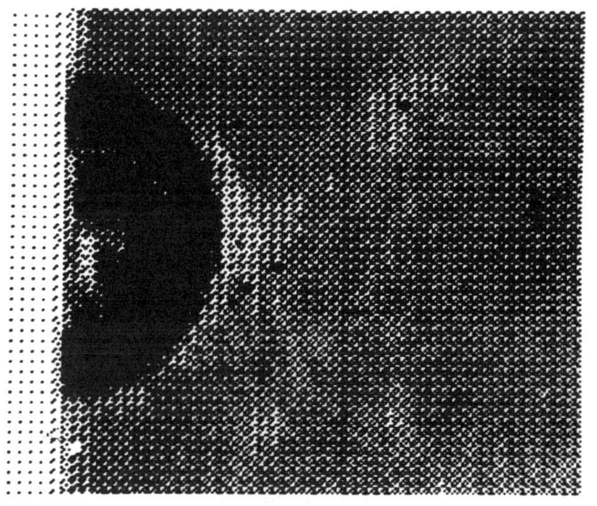

4 × 4 DOT PROFILE A

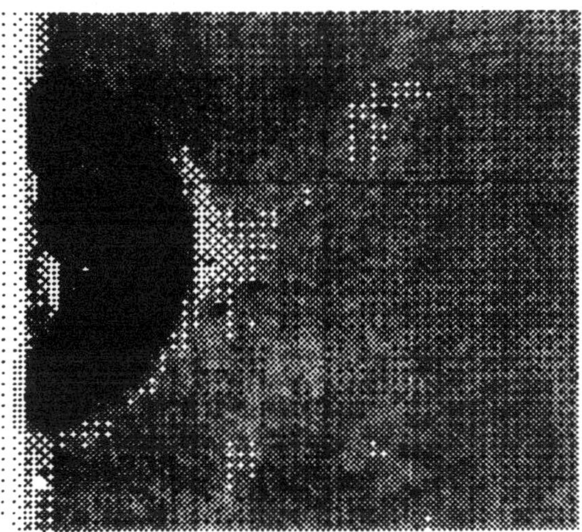

4 × 4 DOT PROFILE B

Region C
Magnification 4X
Scan Line Spacing 0.81 mm (0.032 in)

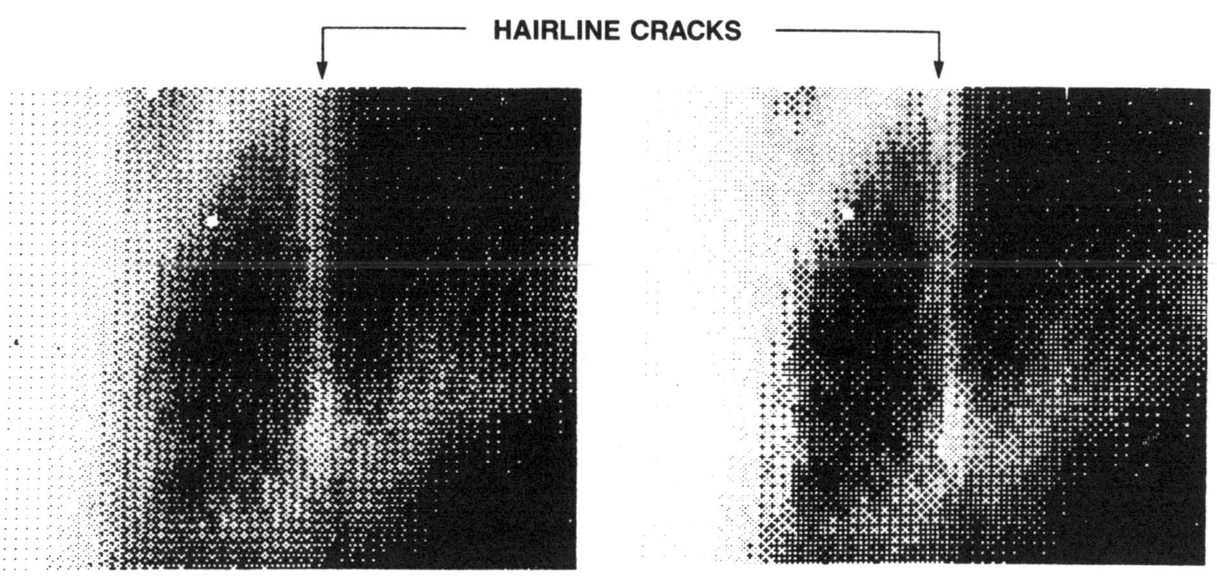

Region D
Magnification 22X
Scan Line Spacing 0.15 mm (0.006 in)

FIGURE 6.2-53 (continued). Ordered dither C-scans of selected regions.

Region E
Magnification 4X
Scan Line Spacing 0.81 mm (0.032 in)

Region F
Magnification 7.2X
Scan Line Spacing 0.46 mm (0.018 in)

FIGURE 6.2-53 (continued). Ordered dither C-scans of selected regions.

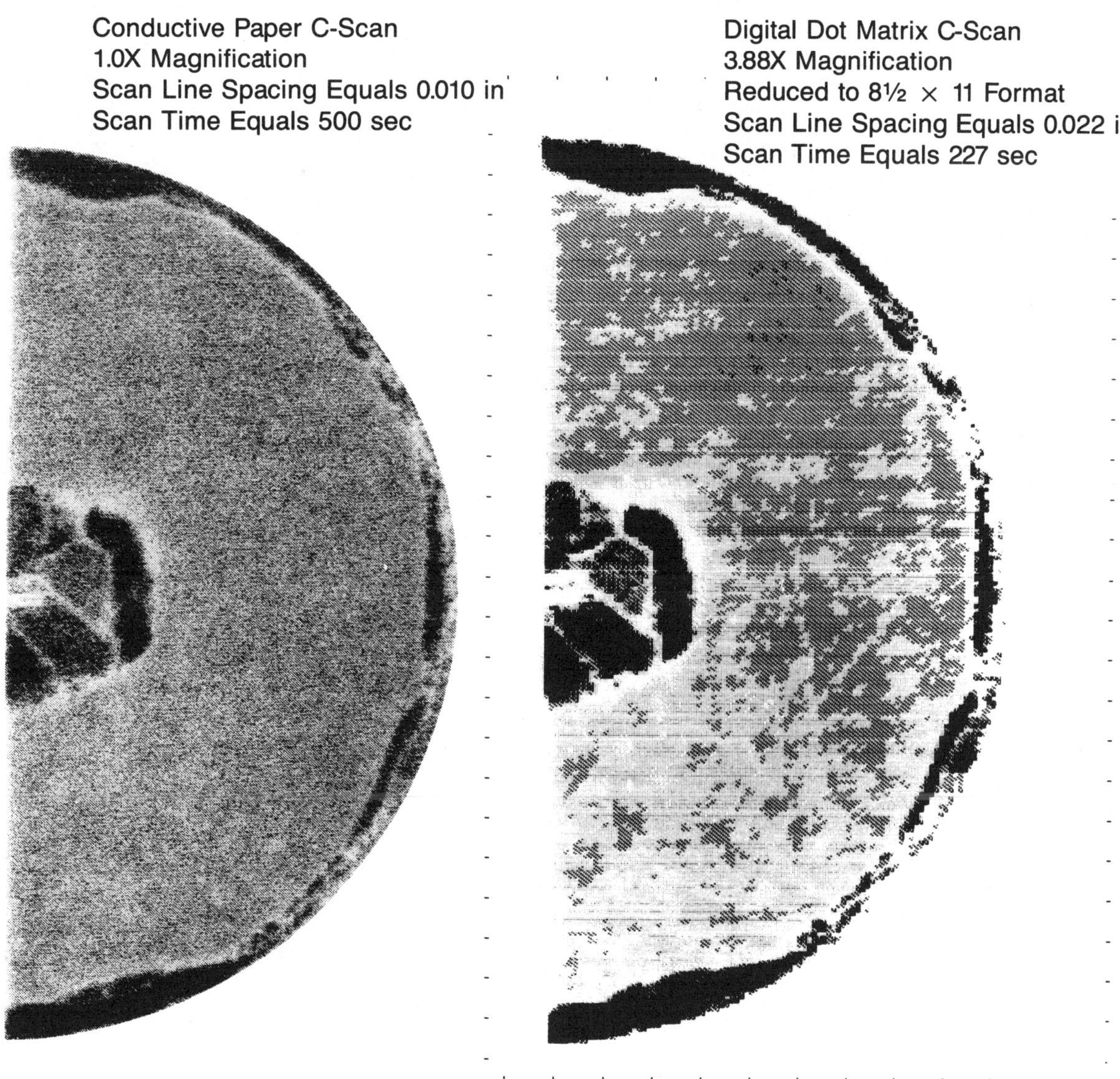

FIGURE 6.2-54. Dot matrix C-scan example.

Ultrasonic C-scans can indicate fiber orientation by the presence of matrix rich areas between plies in continuous fiber laminates.

The structure observed in the woven composite waveform is periodic. The C-scan of the woven composite indicates the woven structure as shown by large fluctuations in the back surface and interlaminar reflections.

HYBRID LAMINATES

Composite materials composed of mixtures of different fiber types within the ply, or changing ply by ply, often allow interrogation only to the first dissimilar material interface. Below is shown a C-scan of a glass/graphite hybrid composed of ¼-inch wide regions of glass fibers uniformly distributed in each graphite ply. Gating was set in the central region of the ultrasonic waveform to display only glass fibers.

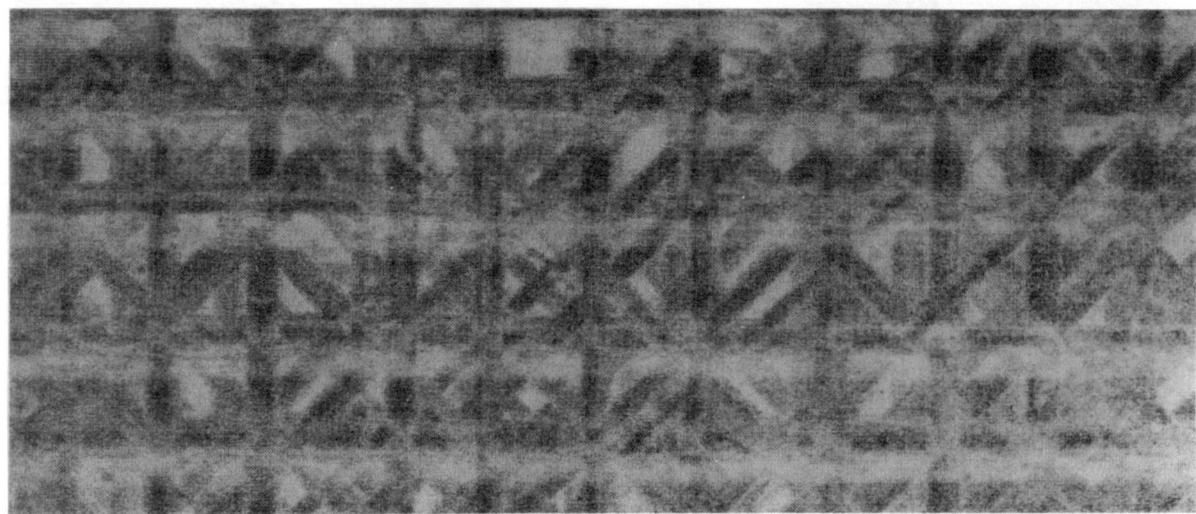

FATIGUE DAMAGE

Fatigue damage in $[45/0/-45/0]_s$ tensile coupons propagates from the edges. The specimens were loaded at S = 67% and R = .1 for 500,000 and 1,000,000 cycles.

500,000 CYCLES **1,000,000 CYCLES**

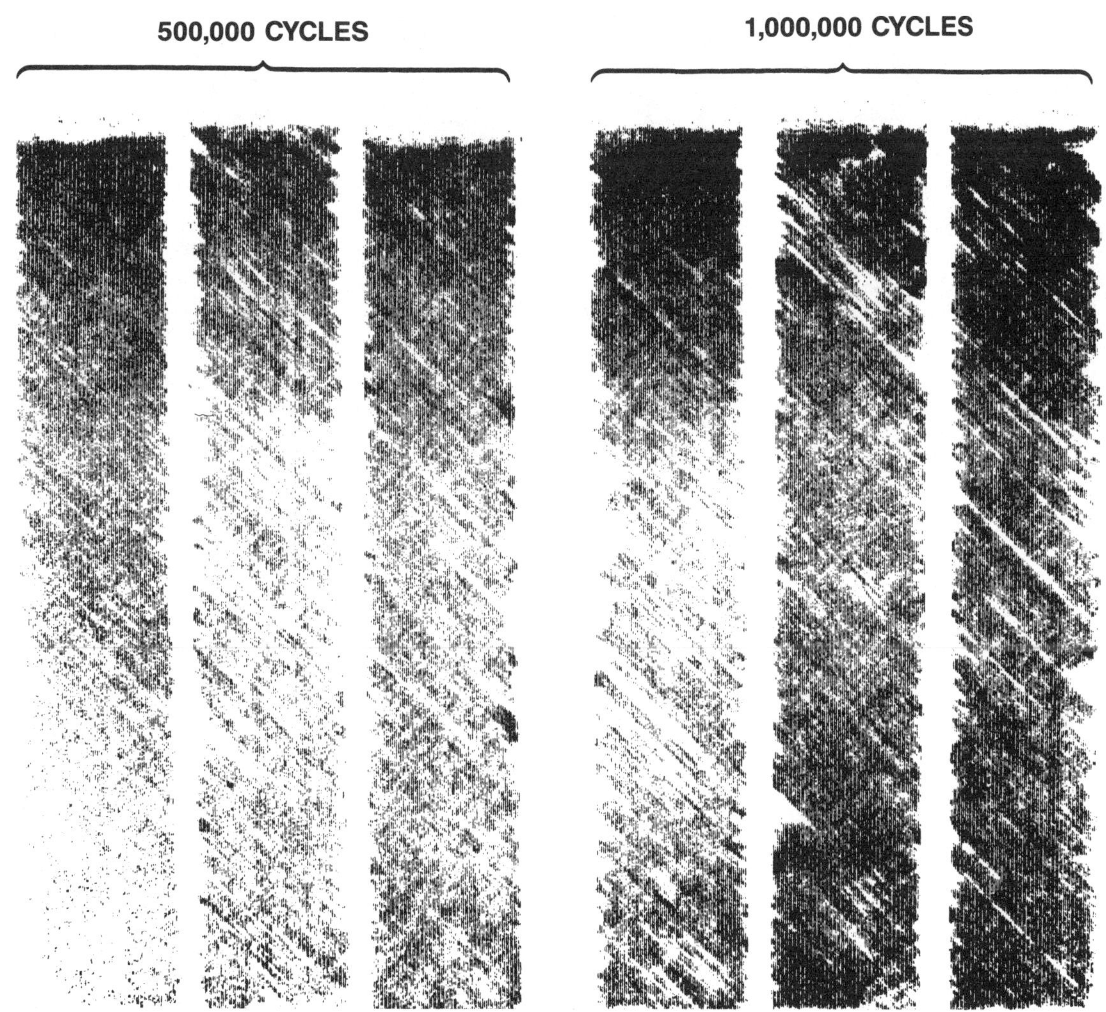

DELAMINATION PROPAGATION

The growth of interlaminar flaws can be monitored using ultrasonic C-scans. Teflon® delaminations were implanted between the mid-plies of a 64 ply, graphite/epoxy, three-point bending specimen. The delamination propagation sequence is shown as a function of the number of fatigue cycles.

S = 0.6   **SPECIMEN 1.5C-4**

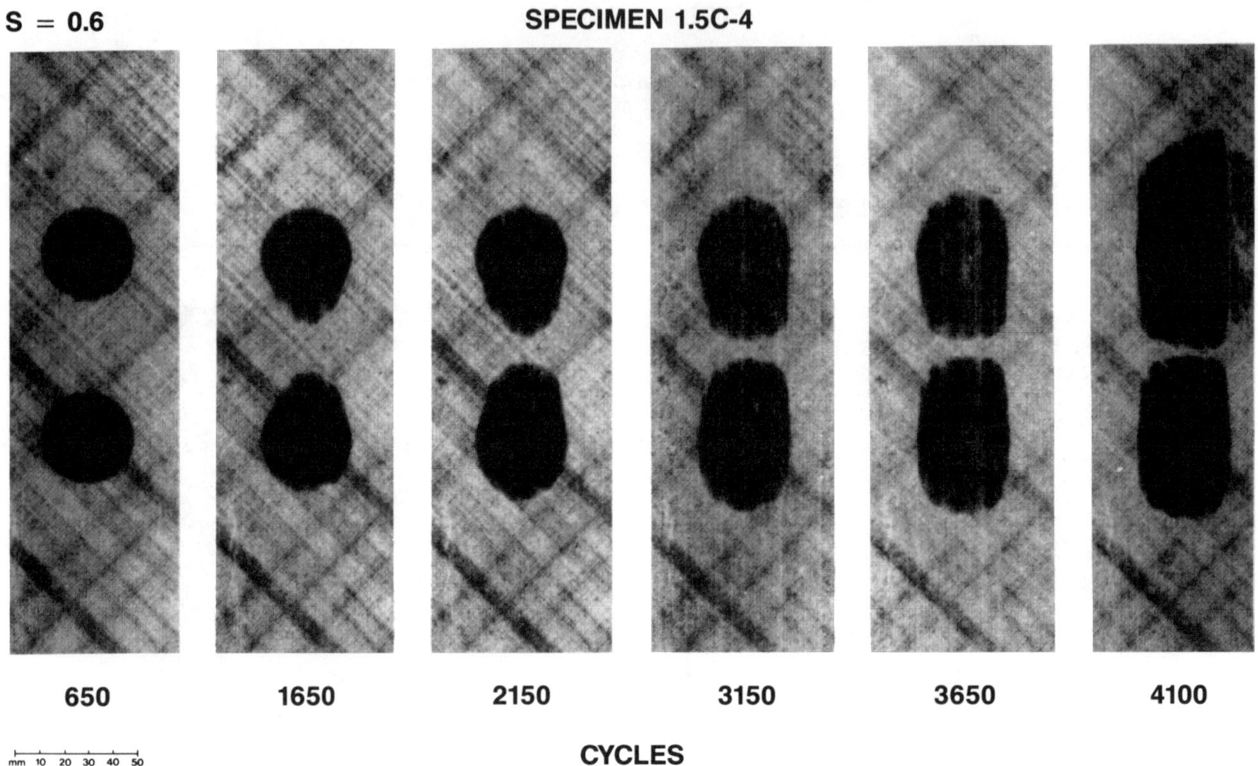

CYCLES

SANDWICH BEAM COMPOSITES

The ultrasonic C-scan of a graphite/epoxy sandwich beam shows the bond between the graphite and the honeycomb structure below.

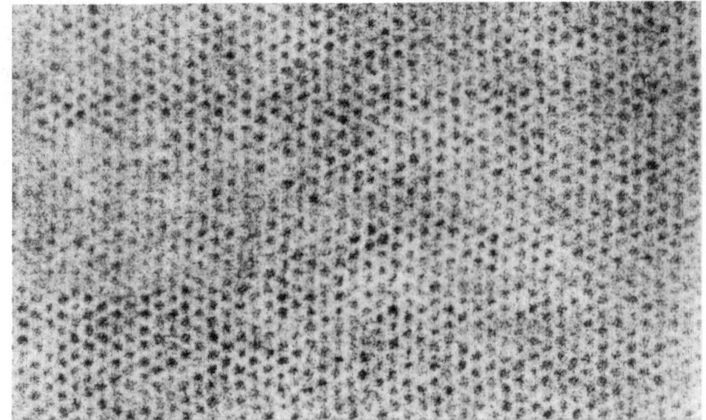

INJECTION MOLDED COMPOSITES

Ultrasonic C-scans indicate flow patterns in short fiber injection molded materials.

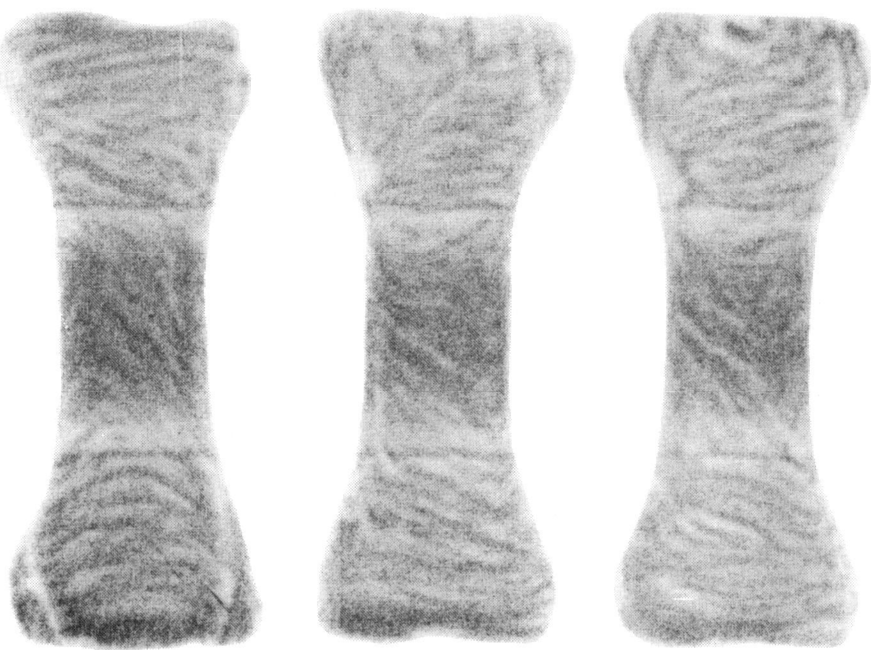

Injection Molded Dog Bone Specimens

INJECTION MOLDED PLATE

This injection molded plate shows a channelling effect and void regions at the end of the mold.

CHANNELLING IN CENTER OF PLATE

VOID REGION

This C-scan of a compression molded plate indicates that six cylinders were compressed to form
the specimen. Knit or weld lines appear where the
cylinders merged.

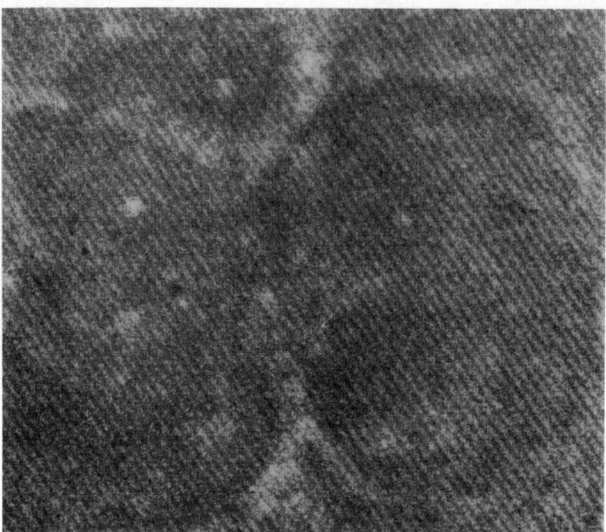

COMPRESSION MOLDING OF DISKS

The gated waveform can be specified to give cumulative information from the back surface reflection, or the gate may be set to capture echoes from the region between the front and back surface
reflections.

Compression Molded Disk
C-Scan with Center Region
Gating Indicating
Particulate Filler Material

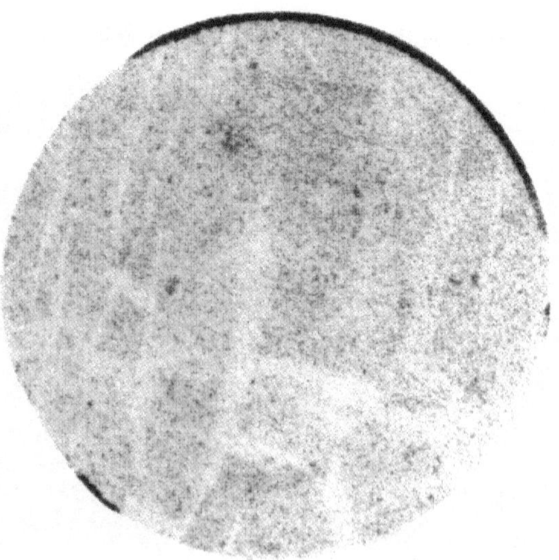

Compression Molded Disk
C-Scan with Back Surface
Gating Indicating Folding
of Molding Compound

106

SMC COMPOSITES

C-scans of low fiber volume fraction sheet molding compounds (SMC) can give an indication of fiber orientation and fiber depleted regions.

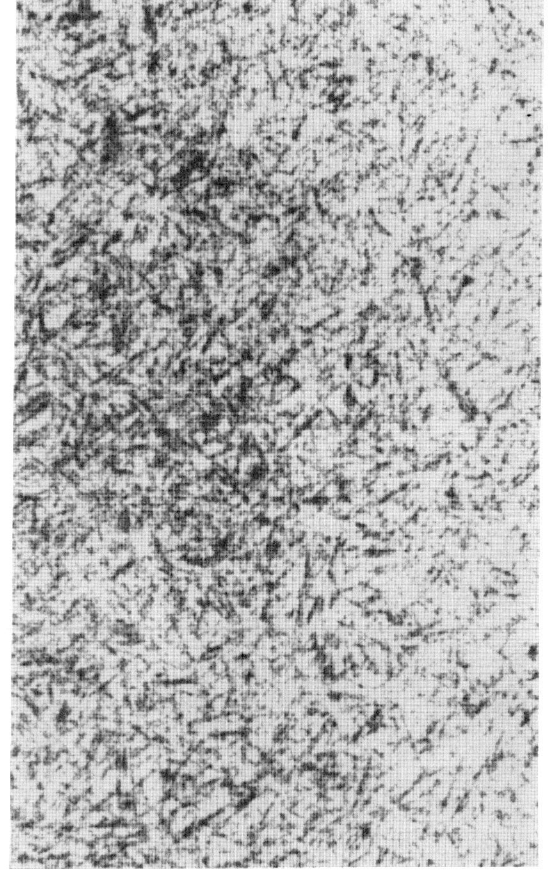

**25% SMC
HOMOGENEOUS REGION**

**25% SMC
FIBER DEPLETED REGION**

HIGH FIBER VOLUME FRACTION SMC

Sheet molding compounds (SMC) with fiber volume fractions of 65% show little fiber orientation in the C-scan; however, void regions are detectable.

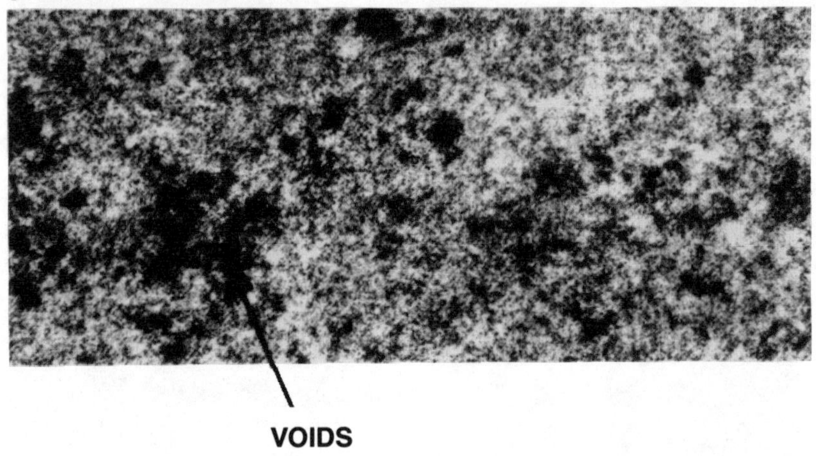

VOIDS

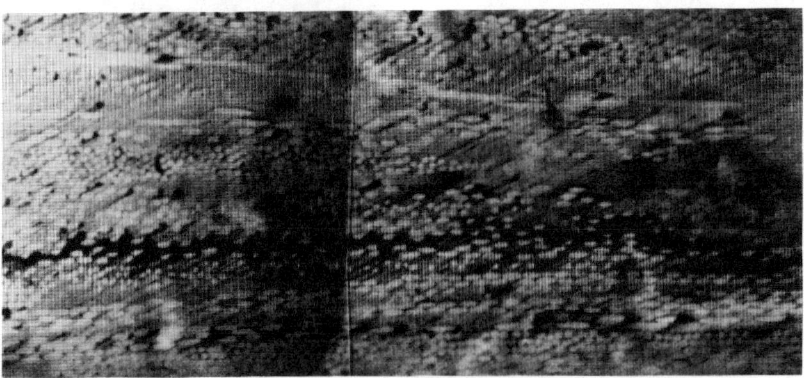

PHOTOMICROGRAPH OF VOID

6.2.7 Appendix: Selected Examples of Composite Material C-Scans

A variety of C-scans of continuous textile and discontinuous fiber reinforced polymeric composite materials are shown on pages 102–108. The examples have been chosen to illustrate the utility of this nondestructive technique to interrogate the composite micro-structure, to assess the quality of the consolidation process, and to identify and locate various types of defects.

6.2.8 References

1. STINCHCOMB, W. W. "Nondestructive Evaluation of Damage Accumulation Process in Composite Laminates," *Composites Science & Tech.*, 25:103–118 (1986).
2. ONO, K. "Acoustic Emission Behavior of Flawed Unidirectional Carbon Fiber-Epoxy Composites," *J. Reinforced Plastics and Composites*, 7:90–107 (1988).
3. MASTERS, J. E. "Correlation of Impact and Delamination Resistance in Interleafed Laminates," *Proc. ICCM & ECCM*, F. L. Matthews et al., eds., Elsevier Applied Science, 3:96–107 (1987).

Interlaminar Fracture of Laminated Composite Materials

6.3.1 Introduction to Interlaminar Fracture Mechanics

Laminated composites are used extensively to replace metallic parts in weight-critical structural applications. A potentially life-limiting failure mode unique to laminated composite structures is their low resistance to interlaminar damage, or "delamination." This type of damage is frequently caused by hard object impact loading, which is a common occurrence during manufacture, service, and maintenance of the structure. The reason for the susceptibility of composites to interlaminar cracking is that there are generally no load-bearing fibers normal to the plane of lamination (i.e., in the thickness direction). The plies are bonded by thin layers of resin that constitute low-energy fracture paths. In contrast to fiber fractures, delaminations have a tendency to grow during cyclic loading [1–3], and this may necessitate inspection procedures in order to maintain the reliability of the structure. Delaminations are especially significant in compression dominated stress fields in the plane of the composite, because instability-related failures may occur [4–6].

A very powerful tool to analyze stresses in solids containing cracks and flaws is fracture mechanics [7]. From an applied mechanics point of view, fracture mechanics aims to predict the onset of fracture for a body containing a crack of given size and geometry. From a materials science point of view, fracture mechanics aims to isolate material parameters of importance to crack resistance so that materials with improved fracture toughness can be devised. It is important to keep these dual purposes in mind when discussing fracture mechanics of composites. Linear elastic fracture mechanics has been found useful for certain types of cracks in composites, i.e., interlaminar cracks. Most of the emphasis to date has been to devise test specimens to characterize fracture behavior and "fracture toughness" of composites. Consequently, the materials science aspects have dominated the field. Applied mechanics has been used extensively to analyze fracture test specimens.

Stress Intensity Factors

The equilibrium of an existing crack may be judged from the intensity of elastic stress around the crack tip or by energy approaches such as the Griffith criterion [8] or the J-integral approach [9]. The elastic stress field in isotropic solids was first considered by Westergaard [10], who approached cracked geometries with complex stress potentials. Sih et al. [11] extended the complex variable approach to derive stress solutions for anisotropic solids.

As an introduction to the analysis of crack problems, the idealizations of plane stress and plane strain in orthotropic materials will be reviewed. For a thin plate with its major dimension and principal material axes in the xy-plane, the relation between strain and stress is approximated by plane stress [12]:

$$\sigma_z = \tau_{yz} = \tau_{xz} = 0 \qquad (6.3\text{-}1)$$

Hooke's generalized law for an orthotropic material with its principal material axes in the xy-plane is [12]

$$
\begin{bmatrix} \epsilon_x \\ \epsilon_y \\ \epsilon_z \\ \gamma_{yz} \\ \gamma_{xz} \\ \gamma_{xy} \end{bmatrix}
=
\begin{bmatrix}
\overline{S}_{11} & \overline{S}_{12} & \overline{S}_{13} & 0 & 0 & \overline{S}_{16} \\
\overline{S}_{12} & \overline{S}_{22} & \overline{S}_{23} & 0 & 0 & \overline{S}_{26} \\
\overline{S}_{13} & \overline{S}_{23} & \overline{S}_{33} & 0 & 0 & \overline{S}_{36} \\
0 & 0 & 0 & \overline{S}_{44} & \overline{S}_{45} & 0 \\
0 & 0 & 0 & \overline{S}_{45} & \overline{S}_{55} & 0 \\
0 & 0 & 0 & 0 & 0 & \overline{S}_{66}
\end{bmatrix}
\begin{bmatrix} \sigma_x \\ \sigma_y \\ \sigma_z \\ \tau_{yz} \\ \tau_{xz} \\ \tau_{xy} \end{bmatrix}
$$

$$(6.3\text{-}2)$$

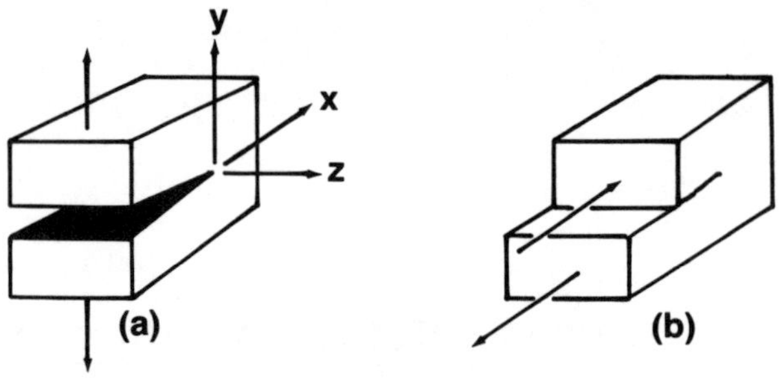

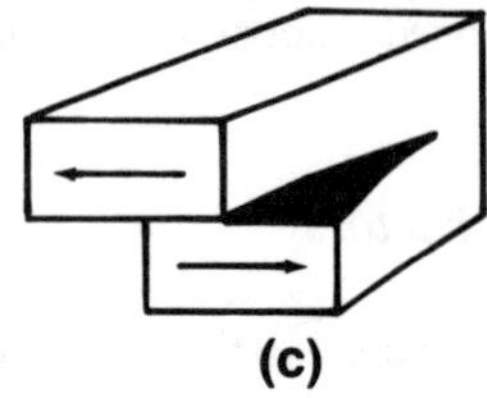

FIGURE 6.3-1. Modes of crack loading: (a) mode I (opening mode); (b) mode II (forward shear); (c) mode III (tearing).

where overbars denote transformed quantities (on-axis to off-axis).

If the stresses according to Equation (6.3-1) are substituted into Equation (6.3-2), it can be simplified to yield

$$
\begin{bmatrix} \epsilon_x \\ \epsilon_y \\ \gamma_{xy} \end{bmatrix} = \begin{bmatrix} \overline{S}_{11} & \overline{S}_{12} & \overline{S}_{16} \\ \overline{S}_{12} & \overline{S}_{22} & \overline{S}_{26} \\ \overline{S}_{16} & \overline{S}_{26} & \overline{S}_{66} \end{bmatrix} \begin{bmatrix} \sigma_x \\ \sigma_y \\ \tau_{xy} \end{bmatrix}
\tag{6.3-3}
$$

Expressions for the compliance elements $\overline{S}_{ij}$ can be found in references [12,13].

For thick plates, conditions of plane strain are commonly assumed by taking

$$
\epsilon_z = \gamma_{yz} = \gamma_{xz} = 0
\tag{6.3-4}
$$

This assumption in combination with Hooke's generalized law [Equation (6.3-2)] leads to

$$
(\overline{S}_{44} - \overline{S}_{45})\tau_{yz} + (\overline{S}_{45} - \overline{S}_{55})\tau_{xz} = 0
\tag{6.3-5}
$$

Since this has to be fulfilled for any combination of material properties, $\tau_{yz} = \tau_{xz} = 0$. Furthermore,

$$
\overline{S}_{13}\sigma_x + \overline{S}_{23}\sigma_y + \overline{S}_{33}\sigma_z + \overline{S}_{36}\tau_{xy} = 0
\tag{6.3-6}
$$

Consequently,

$$
\sigma_z = \frac{-(\overline{S}_{13}\sigma_x + \overline{S}_{23}\sigma_y + \overline{S}_{36}\tau_{xy})}{\overline{S}_{33}}
\tag{6.3-7}
$$

Consequently, σ_z is not an independent quantity and may be removed from Hooke's generalized law (together with τ_{yz} and τ_{xz}) to yield

$$
\begin{bmatrix} \epsilon_x \\ \epsilon_y \\ \gamma_{xy} \end{bmatrix} = \begin{bmatrix} \overline{b}_{11} & \overline{b}_{12} & \overline{b}_{16} \\ \overline{b}_{12} & \overline{b}_{22} & \overline{b}_{26} \\ \overline{b}_{16} & \overline{b}_{26} & \overline{b}_{66} \end{bmatrix} \begin{bmatrix} \sigma_x \\ \sigma_y \\ \tau_{xy} \end{bmatrix}
\tag{6.3-8}
$$

where the constants $\overline{b}_{ij}$ are

$$
\overline{b}_{11} = \frac{\overline{S}_{11}\overline{S}_{33} - \overline{S}_{13}^{\,2}}{\overline{S}_{33}}
\tag{6.3-8a}
$$

$$
\overline{b}_{12} = \frac{\overline{S}_{12}\overline{S}_{33} - \overline{S}_{13}\overline{S}_{23}}{\overline{S}_{33}}
\tag{6.3-8b}
$$

$$
\overline{b}_{16} = \frac{\overline{S}_{16}\overline{S}_{33} - \overline{S}_{13}\overline{S}_{36}}{\overline{S}_{33}}
\tag{6.3-8c}
$$

$$
\overline{b}_{22} = \frac{\overline{S}_{22}\overline{S}_{33} - \overline{S}_{23}^{\,2}}{\overline{S}_{33}}
\tag{6.3-8d}
$$

$$
\overline{b}_{26} = \frac{\overline{S}_{26}\overline{S}_{33} - \overline{S}_{23}\overline{S}_{36}}{\overline{S}_{33}}
\tag{6.3-8e}
$$

$$
\overline{b}_{66} = \frac{\overline{S}_{66}\overline{S}_{33} - \overline{S}_{36}^{\,2}}{\overline{S}_{33}}
\tag{6.3-8f}
$$

For the case in which the xy-system coincides with the principal material axes,

$$
\overline{S}_{16} = \overline{S}_{26} = \overline{S}_{36} = \overline{S}_{45} = 0
\tag{6.3-9a}
$$

and

$$
\overline{b}_{16} = \overline{b}_{26} = 0
\tag{6.3-9b}
$$

As recognized by Sih and Liebowitz [14], Equations (6.3-3) and (6.3-8) are of the same form, and the solution of a plane stress problem can thus be transformed to the corresponding plane strain problem simply by replacing $\overline{S}_{ij}$ with $\overline{b}_{ij}$.

Sih et al. [11,14] investigated elastic stress fields around cracks in bodies under plane stress or plane strain by complex variable methods and singular integral equations for the three basic types of crack surface movements, or fracture modes, illustrated in Figure 6.3-1. These modes are necessary and sufficient to describe all the possible combinations of loads and deformations possible in a cracked body. As the crack tip is approached, the analysis yields stress singularities that are of the order of $r^{-1/2}$, where r is the distance from the crack tip. This result is identical to that obtained in isotropic materials [7], and, as a consequence, stress intensity factors K_I, K_{II}, and K_{III} have been defined in a manner consistent with those for isotropic materials.

As an illustration, consider the anisotropic plate in Figure 6.3-2 loaded by a remote tensile stress of magnitude p. The inclined crack front may be considered as subjected to a far-field tensile stress of magnitude

$$\sigma_\infty = p \sin^2 \alpha \qquad (6.3\text{-}10)$$

and a far-field shear stress of magnitude

$$\tau_\infty = p \sin \alpha \cos \alpha \qquad (6.3\text{-}11)$$

The stress intensity factors K_I and K_{II} may be defined as

$$K_I = \sigma_\infty \sqrt{\pi a} \qquad (6.3\text{-}12a)$$

$$K_{II} = \tau_\infty \sqrt{\pi a} \qquad (6.3\text{-}12b)$$

where a is the semi-crack length. Notice that the stress intensity factors are the same for plane stress and plane strain.

Strain Energy Release Rate

For interlaminar fracture problems, it has become customary to consider the rate of input of work (or strain energy release rate) into the fracture process. The strain energy release rate, G, is a quantity based on energy considerations and is mathematically well defined as well as physically measurable in experiments. The energy approach, which stems from the original Griffith treatment [8], is based on a thermodynamic criterion for fracture by considering the strain energy of the system available for crack growth on one hand and the energy consumed to extend an existing crack on the other hand. In brittle systems, Griffith

found that the dominant source of energy consumption during crack propagation was the generation of additional crack surface energy. In composites, numerous energy absorbing micromechanisms contribute to this macroscopic response: (1) formation of the fracture surface of the main crack; (2) plastic deformation of the matrix in the crack tip region; (3) microcracking and secondary cracks; and (4) crack bridging by fibers resulting in peeling and fracture.

A potential H may be defined for a cracked body as

$$H = W - U \qquad (6.3\text{-}13)$$

where W is the work supplied by the movement of the external forces, and U is the elastic strain energy stored in the body. If G_c is the work required to create a unit crack area, it is possible to formulate a criterion for crack growth:

$$\delta H \geq G_c \delta A \qquad (6.3\text{-}14)$$

where δA is the increase in crack area.

Critical conditions occur when the net energy supplied just balances the energy required, i.e., when

$$\delta H = G_c \delta A \qquad (6.3\text{-}15)$$

The equilibrium becomes unstable when the net energy

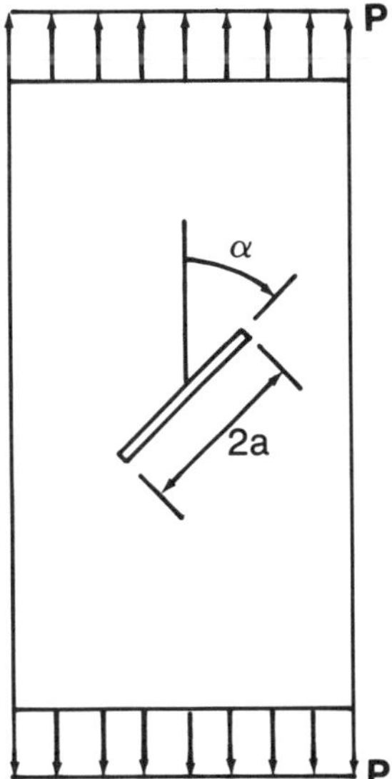

FIGURE 6.3-2. Anisotropic plate containing a crack oriented at an angle α to the loading direction.

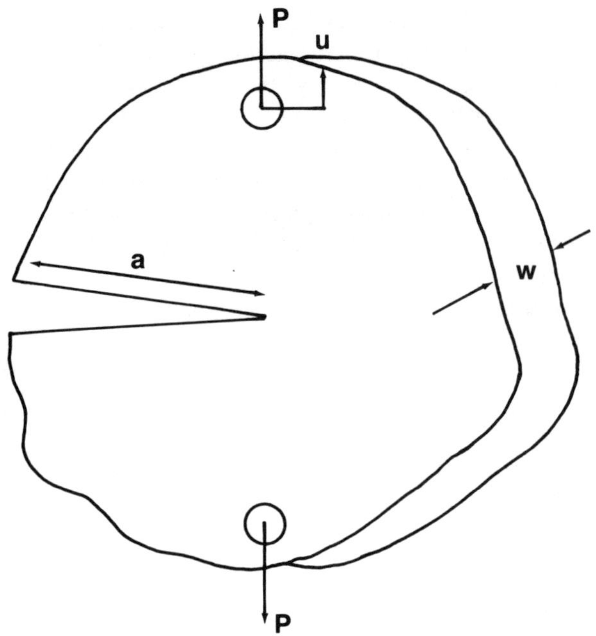

FIGURE 6.3-3. Cracked plate of thickness w, crack length a. Load $= P$ and displacement $= u$.

Consider a cracked plate of width (w) under a load P, as illustrated in Figure 6.3-3. The load-displacement relations for crack lengths a and $a + \delta a$ are illustrated in Figure 6.3-4. As long as there is no crack extension, proportionality exists between displacement and load:

$$u = CP \tag{6.3-20}$$

where C is the compliance of the cracked body ($C = u/P$). The work supplied by the movement of the load P is

$$W = \int_{o}^{u} P\, du \tag{6.3-21}$$

The elastic energy stored in the body is

$$U = \frac{1}{2}\, Pu = \frac{1}{2}\, CP^2 \tag{6.3-22}$$

Now, consider a crack extension from a to $a + \delta a$. Equations (6.3-13), (6.3-19), and (6.3-21) yield

$$G = \frac{1}{w}\left(P\frac{du}{da} - \frac{dU}{da} \right) \tag{6.3-23}$$

Substitution of $u = CP$ into Equation (6.3-23) yields

$$G = \frac{1}{w}\left(P^2\frac{\partial C}{\partial a} + PC\frac{\partial P}{\partial a} - \frac{1}{2}P^2\frac{\partial C}{\partial a} - CP\frac{\partial P}{\partial a} \right) \tag{6.3-24}$$

Consequently,

$$G = \frac{P^2}{2w}\frac{\partial C}{\partial a} \tag{6.3-25}$$

For further illustration of this universal concept, two commonly assumed idealizations, namely constant load (fixed load) and constant displacement (fixed grip) during crack extension will be considered. With consideration to Figure 6.3-4, for the fixed load situation,

$$\delta U = \frac{P\delta u}{2} \tag{6.3-26a}$$

$$\delta W = P\delta u \tag{6.3-26b}$$

supplied exceeds the work of fracture or equivalently the crack growth resistance:

$$\delta H > G_c \delta A \tag{6.3-16}$$

The strain energy release rate, G, is defined as

$$G = \frac{\partial H}{\partial A} \tag{6.3-17}$$

In terms of G, the fracture criterion may thus be formulated as

$$G \geq G_c \tag{6.3-18}$$

Most fracture specimens employed for interlaminar fracture testing are of constant width, w. The amount of new crack surface area, δA, is then $\delta A = w\delta a$, where δa is the incremental crack length that is commonly measured during the fracture experiment. The strain energy release rate can then be expressed as

$$G = \frac{\partial H}{w\partial a} \tag{6.3-19}$$

Equation (6.3-13) gives

$$\delta H = P\delta u - P\delta u/2 = P\delta u/2$$

and Equation (6.3-19) gives

$$G = \frac{P}{2w}\frac{\delta u}{\delta a} = \frac{P^2}{2w}\frac{\partial C}{\partial a} \qquad (6.3\text{-}27)$$

since $u = PC$ and P is a constant in this example. For the fixed grip case, the work term vanishes and

$$\delta U = \frac{u\delta P}{2} \qquad (6.3\text{-}28)$$

Note that δP is negative because of the loss in stiffness followed by crack extension, and G is therefore

$$G = -\frac{u}{2w}\frac{\delta P}{\delta a} = \frac{P^2}{2w}\frac{\partial C}{\partial a} \qquad (6.3\text{-}29)$$

In this case, $P = C/u$ where u is a constant.

Consequently, for both situations, G is equal to the derivative of the elastic energy $Pu/2$ (apart from the sign). Equation (6.3-25) is very convenient for analytical and experimental fracture studies of composites because it relies on the global specimen compliance. As will be shown subsequently, C can be derived from strength of materials principles for many of the interlaminar fracture specimens employed. Alternatively, C can be measured experimentally at various crack lengths and C vs. a relation can be differentiated numerically to obtain the energy release rate G from Equation (6.3-25). This is essentially the basis for experimental compliance calibration methods to be discussed in later subsections.

Fraction Mode Separation. As for the stress intensity factors K_I, K_{II}, and K_{III}, it is possible to separate G into three components:

$$G = G_I + G_{II} + G_{III} \qquad (6.3\text{-}30)$$

Theoretically, the mode separation is based on Irwin's contention that, if the crack extends by a small amount, Δa, the energy absorbed in the process is equal to the

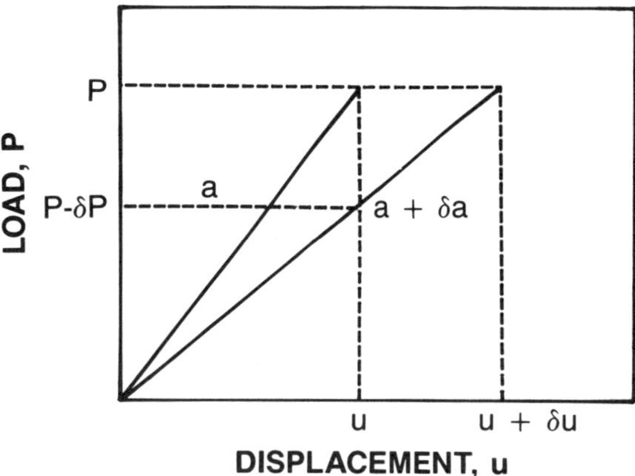

FIGURE 6.3-4. Load-displacement behavior for a cracked body.

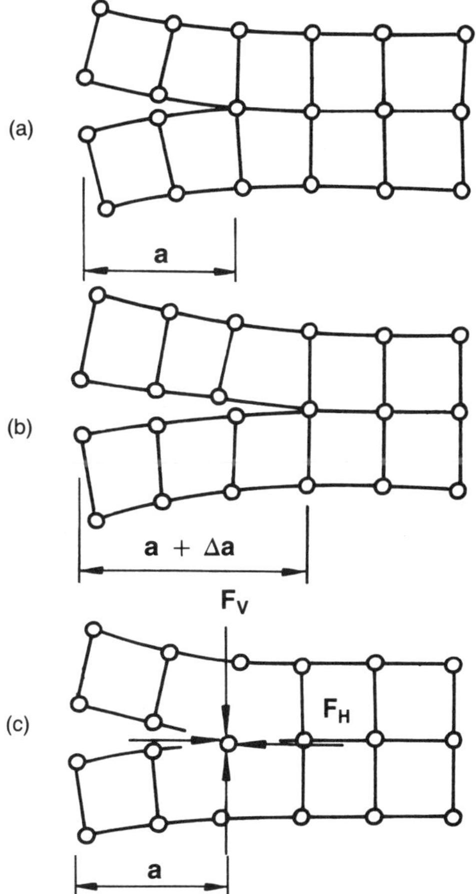

FIGURE 6.3-5. The crack-closure technique in finite elements. F_H and F_V are horizontal and vertical forces applied to close the crack. (a) Original configuration; (b) configuration after release of crack-tip nodes; (c) the released nodes brought back to their initial position by application of nodal forces.

work required to close the crack to its original length [15].

$$G_I = \lim_{\Delta a \to o} \frac{1}{\Delta a} \int_o^{\Delta a} \sigma_y(\Delta a - r,o)\bar{v}(r,\pi)dr \quad (6.3\text{-}31a)$$

$$G_{II} = \lim_{\Delta a \to o} \frac{1}{\Delta a} \int_o^{\Delta a} \tau_{xy}(\Delta a - r,o)\bar{u}(r,\pi)dr \quad (6.3\text{-}31b)$$

$$G_{III} = \lim_{\Delta a \to o} \frac{1}{\Delta a} \int_o^{\Delta a} \tau_{yz}(\Delta a - r,o)\bar{w}(r,\pi)dr \quad (6.3\text{-}31c)$$

where r is the radial distance from the crack tip; σ_y, τ_{xy}, and τ_{yz} are the normal stress, the out-of-plane shear stress, and the in-plane shear stress near the crack tip; and $\bar{v}$, $\bar{u}$, and $\bar{w}$ are the relative opening and sliding displacements between points on the crack faces, respectively.

The crack closure work may also be computed numerically [16]. Figure 6.3-5 illustrates the crack closure method applied to a finite element mesh. A certain crack extension Δa may be introduced in the finite element mesh by releasing duplicate nodes at the crack tip. The resulting deformations are the relative opening and sliding crack tip deformations. By applying nodal forces to the released nodes in three orthogonal directions, it is possible to close the crack tip to its original state. The product of crack tip nodal displacements and forces enables the work of fracture to be evaluated numerically. Figure 6.3-5 illustrates this technique for a mixed mode I and mode II situation. For this situation,

$$G_I \approx \frac{F_V \Delta v}{2w\Delta a} \quad (6.3\text{-}32a)$$

$$G_{II} \approx \frac{F_H \Delta u}{2w\Delta a} \quad (6.3\text{-}32b)$$

F_H and F_V are the horizontal and vertical crack closure forces, and Δu and Δv are the horizontal and vertical increments of displacements required to bring the released nodes to their original positions. Notice that the far-field load applied is constant in the steps illustrated in Figures 6.3-5(a) through 6.3-5(c). Further details on the finite element crack closure technique as applied to anisotropic fracture problems can be found in reference [17].

Relationships between Stress Intensity Factors and Energy Release Rates

As pointed out by Sih and Liebowitz [14], calculation of the components G_I, G_{II}, and G_{III} of the energy release rate, according to Equation (6.3-31), is somewhat academic because cracks do not generally extend in a planar fashion in anisotropic materials. However, if the crack plane coincides with the direction of minimum crack resistance, such as in delamination problems, the direction of crack propagation is collinear with the original crack, and the computed values of G_I, G_{II}, and G_{III} are physically meaningful. In this case, the crack configuration is said to be orthotropic, and the compliance elements defined in Equation (6.3-9) vanish. Substitution of the crack tip stresses and displacements into Equation (6.3-31) yields relationships between energy release rate components and stress intensity factors [11].

The results are

$$G_I = K_I^2 \left(\frac{S_{11}S_{22}}{2} \right)^{1/2} \left[\left(\frac{S_{22}}{S_{11}} \right)^{1/2} + \frac{2S_{12} + S_{66}}{2S_{11}} \right]^{1/2}$$

$$(6.3\text{-}33a)$$

$$G_{II} = K_{II}^2 \frac{S_{11}}{\sqrt{2}} \left[\left(\frac{S_{22}}{S_{11}} \right)^{1/2} + \frac{2S_{12} + S_{66}}{2S_{11}} \right]^{1/2} \quad (6.3\text{-}33b)$$

$$G_{III} = \frac{K_{III}^2}{2(C_{44}C_{55})^{1/2}} \quad (6.3\text{-}33c)$$

where C_{44} and C_{55} are shear stiffnesses defined by

$$\tau_{yz} = C_{44}\gamma_{yz} \quad (6.3\text{-}34a)$$

$$\tau_{xz} = C_{55}\gamma_{xz} \quad (6.3\text{-}34b)$$

The relations for G_I and G_{II}, Equations (6.3-33a) and (6.3-33b), are defined for plane stress. For plane strain, substitute b_{ij} given by Equation (6.3-8) into Equations (6.3-33). It should be recognized that, for highly anisotropic materials, the differences in G_I and G_{II} computed for plane stress and plane strain are small, which may be verified by direct substitution into Equations (6.3-33a) and (6.3-33b).

It may be readily verified that Equation (6.3-33a) applied to isotropic materials yields

$$G_I = \frac{K_I^2}{E} \quad \text{(plane stress)} \qquad (6.3\text{-}35a)$$

$$G_I = \frac{(1 - v^2)K_I^2}{E} \quad \text{(plane strain)} \qquad (6.3\text{-}35b)$$

where E is Young's modulus and v is Poisson's ratio. Similar results are obtained for mode II and mode III. These are the classical relations between stress intensities and energy release rates for isotropic materials [7].

Mixed Mode Fracture

Cracks in isotropic materials tend to grow under mode I conditions [14]. Due to the presence of weak planes between the layers of a composite laminate, however, delaminations are often subjected to a mixed mode stress field. In mixed mode, both normal and shear stresses act across the interface ahead of the crack tip, and both opening and shearing displacements occur on the crack faces behind the interlaminar crack tip. In two-dimensional crack problems involving mode I and II stress intensity factors, fracture toughness characterization must account for the relative amounts of mode I and mode II loadings. Most analyses of mixed mode crack problems employ finite elements [Equation (6.3-32)] to compute the energy release rate components based on nodal forces and displacements. Some analytical schemes for fracture mode separation have been recently proposed by Williams [18] and Suo [19]. These schemes, however, lead to different amounts of mode I and mode II [20], and research in this area is still required.

Another situation where mixed mode crack tip loading may occur is when the crack plane is at the interface between dissimilar isotropic or anisotropic layers [20–25]. In addition to a local mode mix at the crack tip, crack surface interpenetration may occur due to a complex (imaginary) stress intensity factor [21–26]. For commonly employed interlaminar fracture specimens, however, the crack is within a homogeneous orthotropic material, and the aforementioned complexities are avoided. In the remaining subsections, only the three basic fracture modes will be considered.

6.3.2 Mode I Interlaminar Fracture

Interlaminar fracture has aroused considerable attention since the early 1970s, when Pipes and Pagano [27] showed that laminated tensile coupons may fail perpendicular to the direction of load application due to severe interlaminar stresses. Since that time many test methods have been devised to characterize interlaminar failure. Initially, test methods focused on stress criteria for delamination, but later it was widely recognized that fracture mechanics is a more adequate tool to characterize delamination failures, since fracture mechanics accommodates stress singularities at the delamination front. Most of the interlaminar fracture tests are limited to unidirectional laminates where the crack propagates between the plies along the fiber direction. In multidirectional laminates, the crack may have a tendency to branch through the neighboring plies invalidating the coplanar assumption used in the fracture analysis [27].

In this discussion, fracture analysis, testing, and data reduction will be described for the Double Cantilever Beam (DCB) Test for mode I interlaminar fracture characterization of unidirectional composites. The most commonly employed and widely accepted test specimen for mode I interlaminar fracture characterization is the DCB specimen illustrated in Figure 6.3-6. This technique was first employed for composite characterization by Bascom et al. [28], who used a width-

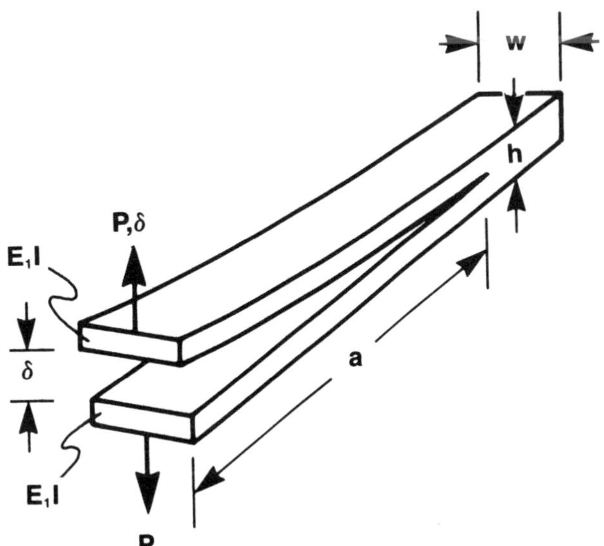

FIGURE 6.3-6. Configuration and test principle of the straight-sided DCB specimen.

tapered specimen. Later, Wilkins et al. [1] proposed a straight-sided DCB specimen that has now become a candidate for standardizing organizations in the U.S., Europe, and Japan [29].

Analysis of the DCB Specimen

The DCB specimen has been the subject for analyses of varying degrees of sophistication, e.g., references [1,18,20,30–33]. In this segment, however, only theories that lend themselves to experimental data reduction will be reviewed. A simple and straightforward analysis of the DCB specimen is the classical beam analysis of Wilkins et al. [1]. It assumes that the loaded beams are rigidly built into the remainder of the uncracked specimen and that classical small deflection beam theory is valid. In this discussion, this theory and an experimental compliance method, initially proposed by Berry [34] to more accurately model the DCB compliance, will be reviewed. A more comprehensive review of analysis methods is presented by Davies and Benzeggagh [35].

Classical Beam Analysis. The compliance C of the DCB specimen may be obtained from elastic beam theory as

$$C = 2a^3/(3E_1I) \qquad (6.3\text{-}36)$$

where a is the crack length and E_1I is the flexural rigidity of each beam of the specimen (see Figure 6.3-6).

The strain energy release rate, G_I, can be obtained from Equation (6.3-25):

$$G_I = \frac{P^2}{2w}\frac{dC}{da} \qquad (6.3\text{-}37)$$

where P is the load applied and w is the specimen width. Equations (6.3-36) and (6.3-37) give

$$G_I = \frac{P^2a^2}{wE_1I} \qquad (6.3\text{-}38)$$

Critical conditions occur when $P = P_c$, i.e., when the mode I interlaminar fracture toughness, G_{IC}, is

$$G_{IC} = \frac{P_c^2a^2}{wE_1I} \qquad (6.3\text{-}39)$$

Notice that this analysis assumes plane stress conditions.

Experimental Compliance Method. According to the Berry [34] approach, the beam compliance is expressed as

$$C = \frac{a^n}{H} \qquad (6.3\text{-}40)$$

where n and H are parameters to be determined experimentally. Comparison with the classical beam Equation (6.3-37) shows that the two approaches coincide if $n = 3$ and $H = 3E_1$, $I/2$. The critical value of G_{IC} according to this approach is

$$G_{IC} = \frac{nP_c^2a^{n-1}}{2wH} \qquad (6.3\text{-}41)$$

Substitution of $\delta_c = P_c a^n/H$, where δ_c is the displacement at the critical load, yields

$$G_{IC} = \frac{nP_c\delta_c}{2wa} \qquad (6.3\text{-}42)$$

The Berry method is currently accepted by ASTM as the basis for experimental fracture toughness determination of composites [36]. The reader should be alerted, however, that alternative analyses for reduction of DCB test results are considered by the European Group on Fracture (EGF), ASTM, and Japanese Industrial Standards (JIS) [37]. At the present time, the Berry method is considered appropriate, and its application to experimental DCB test results will be outlined in a subsequent discussion. Comparison to the classical beam method and a direct method, called the *area method*, will also be presented. Prior to description of data reduction methodology, specimen design and testing issues will be outlined.

DCB Specimen Design

The classical beam expressions and their empirical extensions for the DCB compliance and energy release rate do not incorporate interlaminar shear deformation and large displacement response. As discussed by Devitt et al. [38], large displacements of the beams of the DCB specimen can, in principle, be accounted for, but the data reduction becomes much more involved. A similar reasoning applies to the case when shear deformation is significant. For straightforward and accurate determination of G_{IC}, the DCB specimen should be sized so that the influences of interlaminar shear deformation and large deflections are minimized. Such an analysis was presented by Wilson and Carlsson [39], and the results are reproduced here.

Interlaminar shear deformation in the beams of the DCB specimen can be minimized by considering the additional beam deflection due to interlaminar shear. The deflection of each beam, Δ, can be expressed [39] as

$$\Delta = \Delta_B \left(1 + \frac{3h^2 E_1}{40a^2 G_{13}} \right) \qquad (6.3\text{-}43)$$

where Δ_B is the beam deflection neglecting interlaminar shear deformation, h is the total DCB thickness, and a is the crack length (see Figure 6.3-6). G_{13} is the interlaminar shear modulus in the 1–3 plane.

By Equation (6.3-43), the influence of shear deformation can be minimized to a certain fraction, f_s:

$$f_s = \frac{3h^2}{40a^2} \left(\frac{E_1}{G_{13}} \right) \qquad (6.3\text{-}44)$$

Assuming $f_s = 0.05$, Equation (6.3-44) yields

$$\frac{a}{h} \geq \sqrt{\frac{3(E_1/G_{13})}{2}} \qquad (6.3\text{-}45)$$

For a typical graphite/epoxy material, E_1/G_{13} is approximately 25 [40] and Equation (6.3-45) yields $a/h \geq 6.1$. This means that for a twenty-four-ply laminate the minimum crack length should be at least 20 mm.

An upper bound on crack length, based on the requirement of small deflection, linear elastic load-displacement response [39], is given by

$$a \geq 9.2 \times 10^{-2} \sqrt{\frac{E_1 h^3}{G_{Ic}}} \qquad (6.3\text{-}46)$$

For a brittle graphite/epoxy composite with $E_1 = 140$ GPa and $G_{Ic} = 0.19$ kJ/m², the maximum permissible crack length is 42 cm, well beyond that required for most specimens.

DCB Testing Issues

Load tabs are required in order to transfer tensile load to the beams. Several tab configurations have been proposed, such as "T-tabs" cut from extruded aluminum, "end blocks," and "piano hinges" (see references

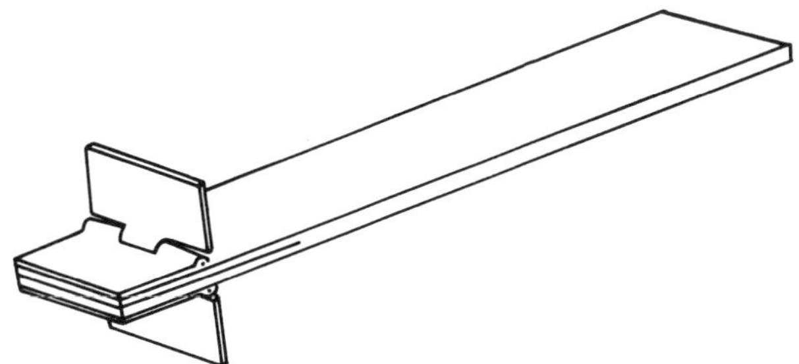

FIGURE 6.3-7. Hinged DCB specimen.

[36,37]). The piano hinge configuration shown in Figure 6.3-7 has been recommended as the ASTM standard [36]. The DCB specimen should be 15 cm long and 2.5 cm wide. To provide a starter crack, thin inert films, e.g., PTFE and polyimide, should be coated with a release agent and placed at the mid-plane of the laminate. These films are typically 15–75 μm in thickness. The minimum film thickness should be used in order to achieve a naturally occurring delamination in a composite structure. Too small film thickness may, on the other hand, lead to wrinkling of the insert during processing, which leads to a thick resin rich zone between the plies around the insert (see Figure 6.3-8). Too large film thickness may also lead to a thick resin rich zone at the crack tip. A typical graphite/epoxy DCB specimen employs twenty-four plies. Thus, the film starter is placed between the twelfth and thirteenth plies. ASTM [41] recommends a 50 mm-long precrack (measured from the point of load introduction) and a hinge length (bonded length in Figure 6.3-7) of 12.5 mm. Procedures for surface preparation of the DCB specimen and hinges in order to achieve adequate bonding are provided in references [41,42]. Hinge alignment is critical and may be accomplished by using a mounting jig [42] or other procedures.

Specimens should be conditioned to contain a uniform moisture content by drying for 10 days at 77°C. Drying should be performed after the hinges are

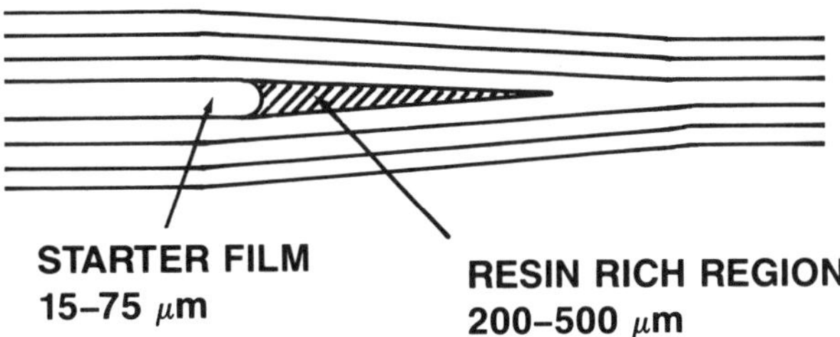

FIGURE 6.3-8. Resin rich region ahead starter film [35].

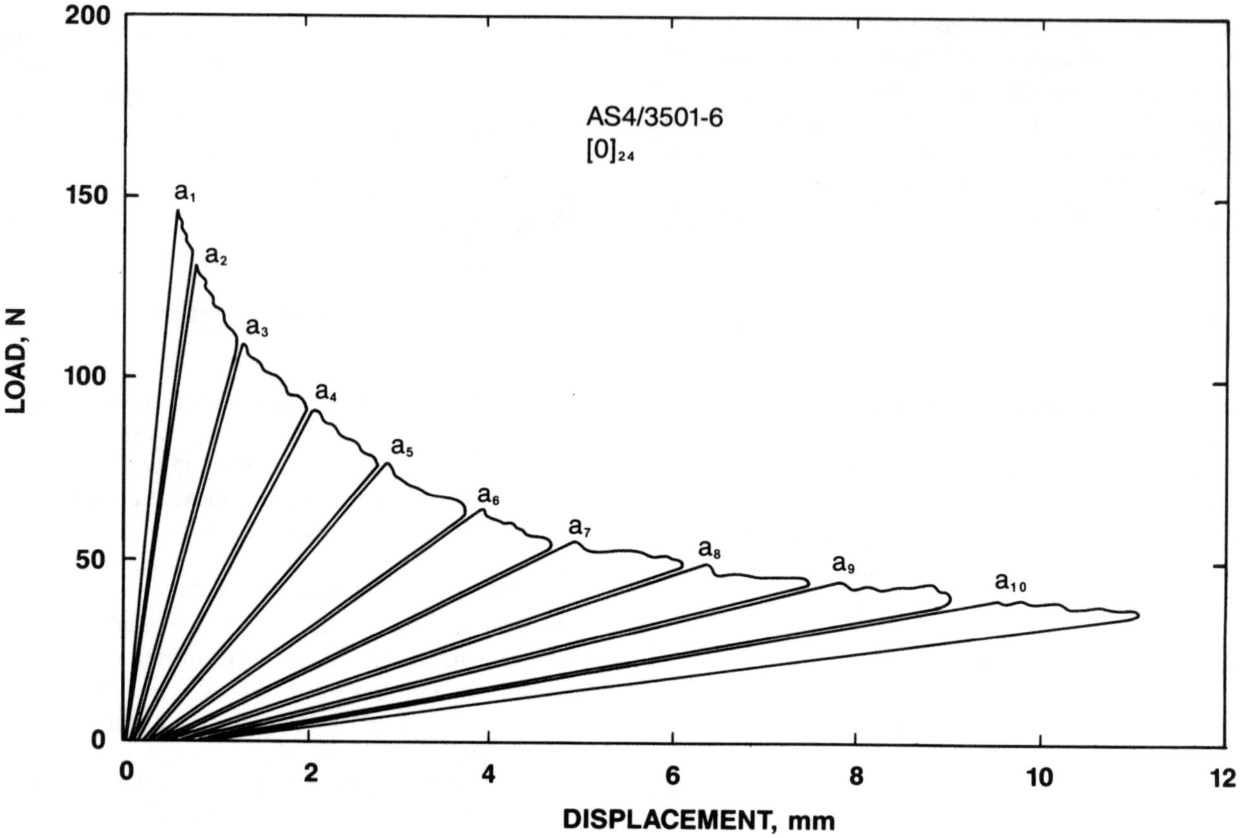

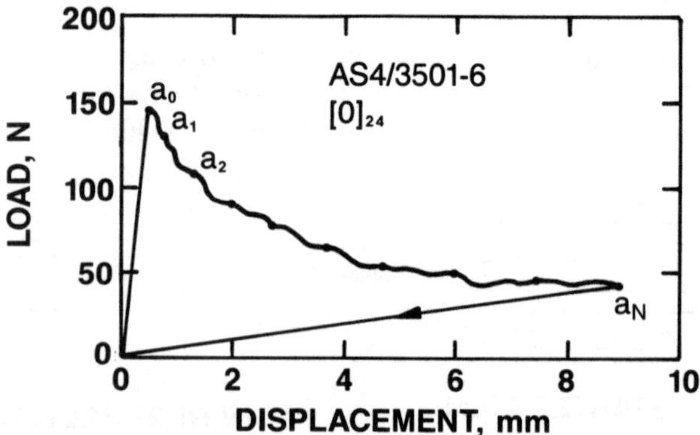

FIGURE 6.3-9. Load-displacement curves for a $[0]_{24}$ graphite/epoxy DCB specimen at various crack lengths.

bonded to the specimen to take advantage of any post-curing of the adhesive that may occur during the drying cycle. Following the drying cycle, specimens may be stored in a dessicator for up to one day and then removed immediately before testing. Measure the width and thickness of each specimen to the nearest 0.05 mm,

FIGURE 6.3-10. Load-displacement record for continuous loading of a DCB specimen.

at the mid-point, and at 25 mm from either end. The variation in thickness along the length of the specimen should not exceed 0.1 mm. Average values of the width and thickness measurements should be recorded.

Mark the edges of the specimen at 2.5 mm increments up to 7.5 cm beyond the starter crack with a thin coating of brittle typewriter correction fluid, or equivalent, for visual crack determinations.

DCB Testing. The test procedure employed depends on the data reduction procedure used. While the classical beam approach by Wilkins et al. [1] and the area method (to be described later) require loading and unloading cycles for each increment of crack growth, the proposed standards [37,41] require continuous loading of the DCB specimen. Each method has its own merits that have to be considered prior to selecting one method. In this segment both types of tests will be described. Irrespective of test method chosen, a properly calibrated and aligned load frame should be used.

For crack length measurement, an optical microscope or a precision dial caliper is needed. Determine the initial crack length, a, from the loading line to the tip of the starter crack on both sides of the specimen. Use a cross-head rate of about 0.5 mm/min for loading the specimen. A real-time analog display of the load versus crack opening displacement curve should be generated during the test on an x-y recorder. Displacements can be determined from an extensometer or LVDT (linear variable differential transformer) attached to the specimen or from the cross-head travel if machine compliance is negligible.

The classical beam approach by Wilkins et al. [1] and the area method, to be described later, require loading-unloading cycles. For these methods, the DCB specimen is loaded until the crack extends about 10 mm and the cross-head is stopped. At this point, the actual crack length is measured and the specimen unloaded. Mark the crack length on the recorder chart for subsequent identification. Repeat this procedure until the crack is approximately 100 mm in length. Figure 6.3-9 shows typical results for a $[0]_{24}$ graphite/epoxy DCB specimen.

For the continuous loading dictated by the ASTM test procedure [41], the test is conducted according to Figure 6.3-10. As the critical load is reached, the crack starts to grow. Mark the crack length on the chart when the crack reaches the first marking (which is 2.5 mm ahead of the 50 mm starter crack). Continue this procedure until the crack advances about 40 mm beyond the initial crack tip. At this point, the total crack length is about 90 mm.

DCB Data Reduction Methodology

For the discontinuous load-displacement data in Figure 6.3-9, two common data reduction methods, namely the classical beam theory method by Wilkins et al. [1] and the area method by Whitney et al. [43], will be reviewed.

In the classical beam theory method [1], the compliance, C, is evaluated from the rising portions of each load cycle ($C = \delta/P$). Plot C versus a in log-log form as shown in Figure 6.3-11. Force-fit a straight line with a slope of 3.0 to the data by linear regression analysis or by hand. Extrapolate the line to log $a = 0$ ($a = 1$ mm). The corresponding compliance is, according to Equation (6.3-36), equal to $2/(3E_1I)$. In order to obtain the mode I fracture toughness, G_{IC}, the critical loads, P_c, at the onset of crack propagation are evaluated from the load versus displacement record (Figure 6.3-9). The P_c versus a data are plotted in a log-log diagram, and a

line with a slope equal to -1 is fit to the data. Extrapolate the line to $a = 1$ mm. The critical load at $a = 1$ mm corresponds to the constant $\sqrt{wE_1IG_{IC}}$ according to Equation (6.3-38). The two constants so determined are hence

$$A_1 = \frac{2}{3E_1I} \tag{6.3-47a}$$

$$A_2 = \sqrt{wE_1IG_{IC}} \tag{6.3-47b}$$

Equations (6.3-47a) and (6.3-47b) combined yield

$$G_{IC} = \frac{3A_1A_2^2}{2w} \tag{6.3-48}$$

As suggested in reference [42], it is sometimes more practical to extrapolate the curves to $a = 100$ mm than extrapolating them to $a = 1$ mm. Extrapolating to

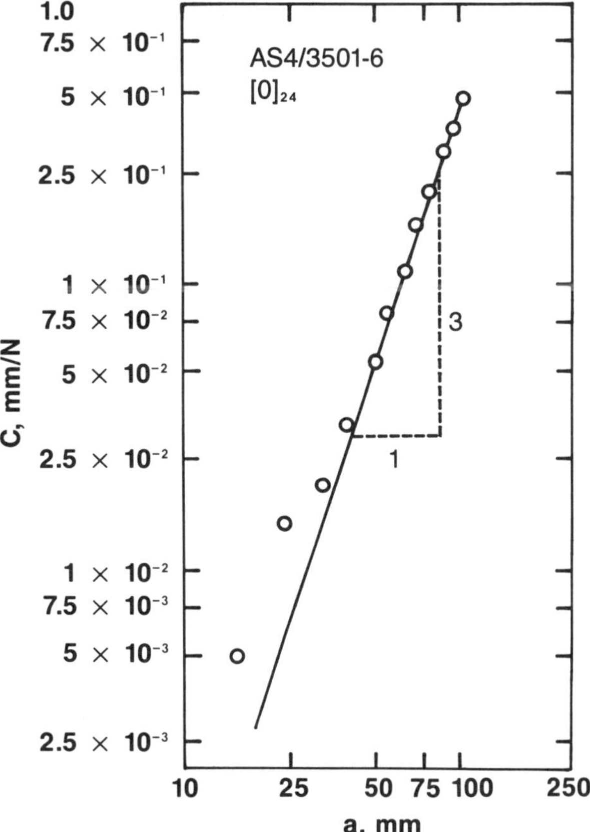

FIGURE 6.3-11. Compliance versus crack length relation for a DCB specimen.

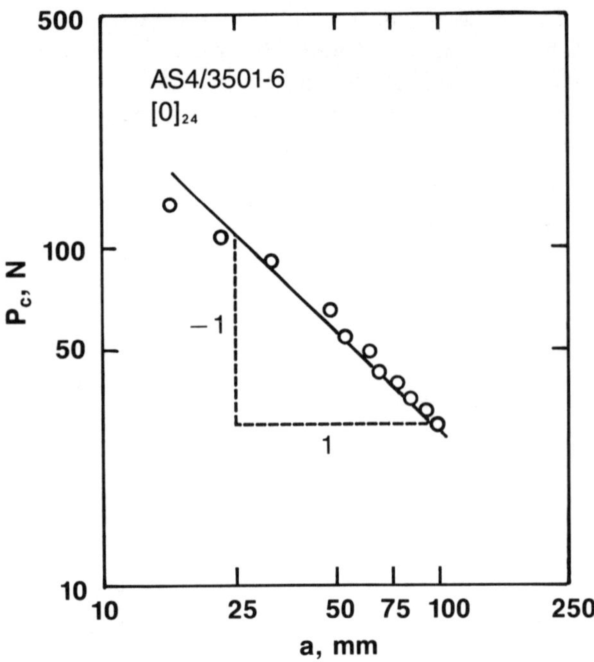

FIGURE 6.3-12. Critical load versus crack length for a DCB specimen.

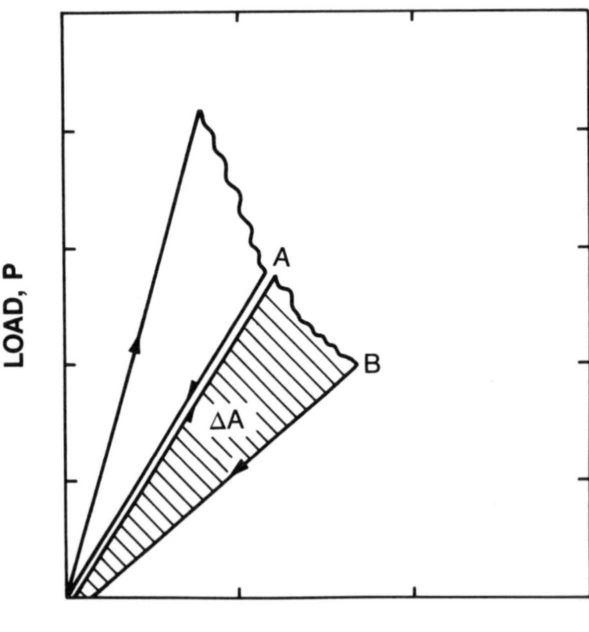

FIGURE 6.3-13. Area method to determine G_{IC}.

$a = 100$ mm yields C (100) and P_c (100). A_1 and A_2 become $A_1 = 10^{-6}C$ (100) and $A_2 = 100 P_c$ (100). This method applied to the graphite/epoxy data of Figures 6.3-11 and 6.3-12 yields $G_{Ic} = 190$ J/m². This method provides an averaged G_{Ic} value for the entire range of crack lengths, which is appropriate if G_{Ic} is a true constant, independent of crack extension. As will be shown subsequently, however, G_{Ic} may tend to increase with crack length, which may not be captured by this method. In order to capture the increasing tendency of G_{Ic} with crack extension (*R*-curve effect), Equations (6.3-39) and (6.3-47a) may be combined to yield

$$G_{IC} = \frac{3A_1 P_c^2 a^2}{2w} \qquad (6.3\text{-}49)$$

A_1 is obtained from Figure 6.3-11, but the curve fit procedure of Figure 6.3-12, which assumes a constant G_{IC} value, is circumvented. By this approach, individual sets of P_c and a values are substituted into Equation (6.3-49) to extract a possible crack length dependency to G_{IC}. The results of this method will be illustrated subsequently.

Area Method. An alternative method to determine the fracture toughness is the area method, which allows for the direct evaluation of G_{IC}. The critical strain energy release rate may be determined from a loading-unloading sequence according to Figure 6.3-13. From the definition of the strain energy release rate [Equation (6.3-19)], G_{IC} may be obtained as

$$G_{IC} = \frac{\Delta A}{w(a_2 - a_1)} \qquad (6.3\text{-}50)$$

where ΔA is the area indicated in Figure 6.3-13, and $a_2 - a_1$ is the increment in crack length.

For linear elastic behavior, G_{IC} is approximately [43]

$$G_{IC} = \frac{P_A \delta_B - P_B \delta_A}{2w(a_2 - a_1)} \qquad (6.3\text{-}51)$$

where P_A and δ_A are the load and displacement, respectively, at point A in Figure 6.3-13. P_B and δ_B are the corresponding quantities at point B. An average G_{IC} value is obtained from the total series of loading and unloading curves in Figure 6.3-9. One of the major advantages of the area method is that it quantifies the propagation toughness while the compliance techniques characterize the initiation toughness. In some materials, stick-slip behavior is observed [64] where unstable growth occurs, even though the DCB specimen promotes inherently stable crack growth. In this case, propagation toughness as measured by the area method would be less than the initiation toughness.

The area method was applied to the results shown in Figure 6.3-9. The average value of G_{IC} was determined to be 0.21 kJ/m². This value is relatively close to G_{IC} determined from the compliance method ($G_{IC} = 0.19$ kJ/m²), since stable growth is observed in this graphite/epoxy composite.

Experimental Compliance Method. Fracture toughness determination according to this method requires the parameter n [see Equation (6.3-42)] to be determined from experimental measurements. n is obtained as the slope of a plot of P_c/δ_c versus a in a log-log diagram as illustrated in Figure 6.3-14. The parameter n is obtained as the magnitude of the slope of the line (see Figure 6.3-14). This may easily be verified from Equation (6.3-40). Once n is obtained, the fracture toughness is determined for *each crack length* from

$$G_{IC} = \frac{n}{2w} \frac{P_c \delta_c}{a} \qquad (6.3\text{-}52)$$

In this equation, P_c, δ_c, and a are the critical load, critical displacement, and crack length corresponding to the ith mark on the curve in Figure 6.3-10. This procedure thus provides G_{IC} for each crack length.

Figure 6.3-15 shows G_{IC} determined from the classical beam theory method [Equation (6.3-49)] and the experimental compliance method [Equation (6.3-52)]. It is observed that the experimental compliance method is slightly conservative and that G_{IC} increases with crack length to an apparent equilibrium value (*R*-curve effect). Reasons for the *R*-curve will be discussed in the following segment.

R-*Curve Effect*

A complicating feature in mode I interlaminar fracture testing is the increase in resistance with crack extension. This behavior is largely attributed to pull-out and fracture of bridged fibers during the mode I fracture process (see Figure 6.3-16).

When the DCB specimen is loaded, the bridged fibers will be pulled out of the matrix, or they will fracture [44]. These processes obviously contribute to the

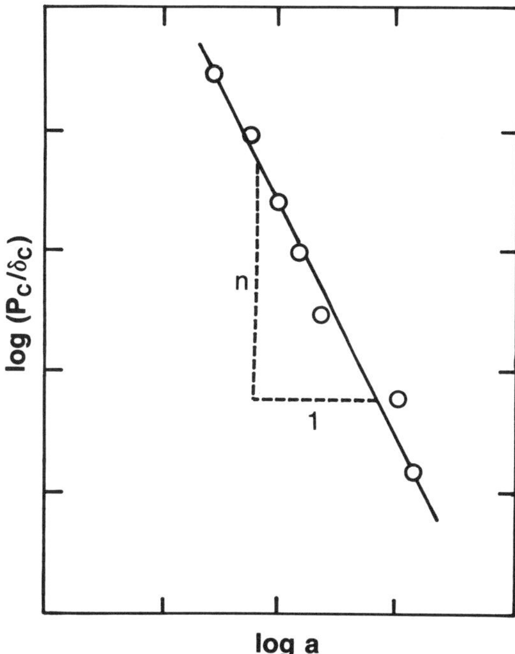

FIGURE 6.3-14. Plot used to determine parameter n.

work performed in separating the fracture surfaces [45–47] and lead to a larger apparent fracture toughness value, G_{IC}. It should be pointed out that fiber bridging is a phenomenon restricted to the unidirectional lay-ups commonly used to characterize interlaminar fracture toughness. It is generally attributed to "nesting" of fibers from adjacent plies. During the pressure/temperature cure cycle employed in processing a panel, fibers on adjacent plies of the crack inter-

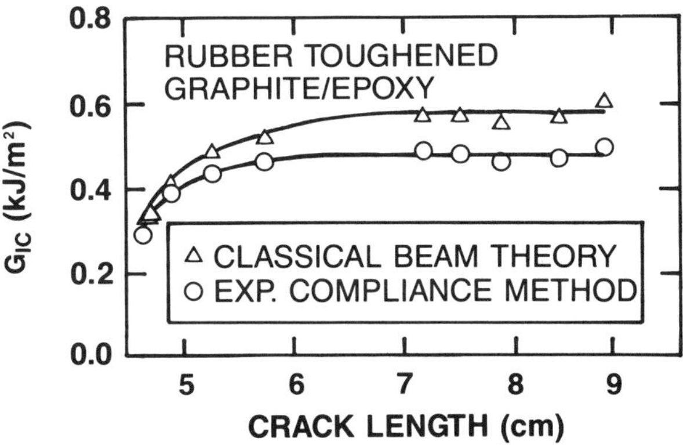

FIGURE 6.3-15. G_{IC} determined from classical beam theory and experimental compliance calibration versus crack length.

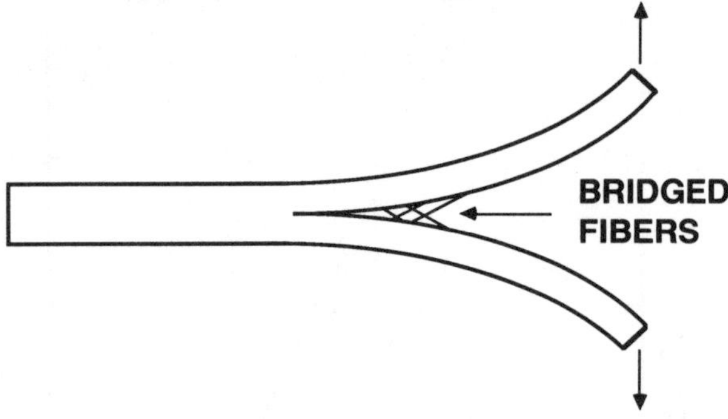

FIGURE 6.3-16. Schematic of fiber bridging in DCB testing.

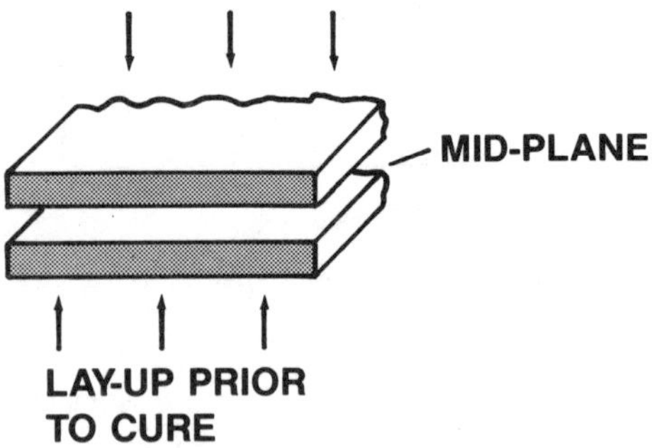

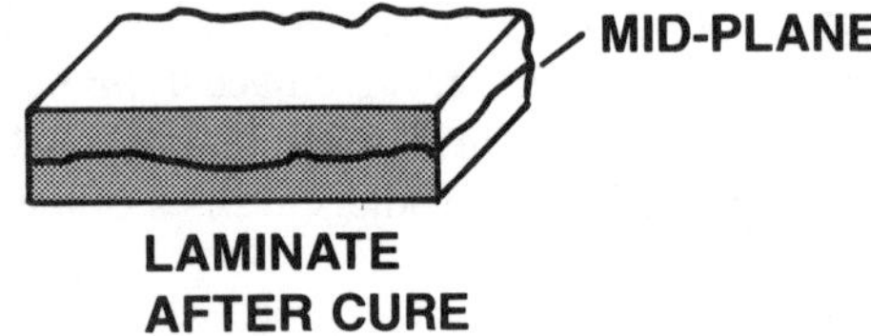

FIGURE 6.3-17. Fiber nesting at the mid-plane of a unidirectional laminate.

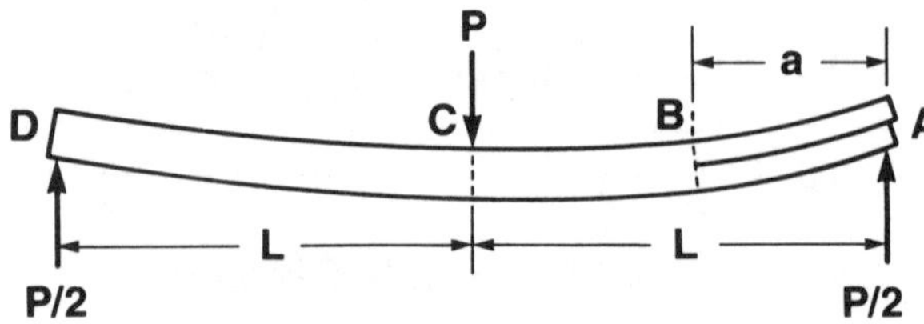

FIGURE 6.3-18. End Notch Flexure (ENF) specimen. Total span is $2L$, total thickness is $2h$, and the width is w.

face may migrate, causing a non-unique delamination plane (see Figure 6.3-17). During subsequent delamination growth the delamination may wander among the fibers, and fiber bridging occurs.

To explain the influence of fiber bridging on delamination toughness, it is necessary to examine the fracture test in more detail. In order to define a site for delamination propagation, a thin film insert is placed at the mid-plane of the laminate prior to processing. At the first loading of the DCB specimen, when the delamination grows from the film insert starter crack, there is no fiber bridging. As the crack opening displacement is increased during subsequent crack growth increments, fiber bridging develops in many composite systems, causing the apparent fracture toughness to rapidly increase with crack extension until an equilibrium density (number of bridged fibers per unit area) is reached [44,47] (see Figure 6.3-15). The R-curve may be characterized by an initiation value, G_{IC} (init) and a steady state propagation value, G_{IC} (prop).

As demonstrated by Hashemi et al. [47], the initiation toughness, G_{IC} (init), is dependent on the amount of precracking beyond the crack starter film. To obtain the full R-curve, Hashemi et al. [47] showed that the extent of precracking (cracking growth beyond the tip of the insert film) should be kept at a minimum. To propagate the delamination directly from the starter crack may lead to an erroneous overestimate of the initiation toughness, G_{IC} (init), due to a resin rich region at the tip of the film insert [47] (see Figure 6.3-8). The increase in fracture toughness with crack extension illustrated in Figure 6.3-15 (besides fiber bridging) may also be due to an increase in the degree of plastic deformation and microcracking around the crack tip with crack extension [47].

This discussion shows that, in order to fully characterize the mode I fracture behavior, the R-curve should be established. Nevertheless, many studies fail to report the initiation toughness, G_{IC} (init), and report only the steady state propagation toughness, G_{IC} (prop).

6.3.3 Mode II Interlaminar Fracture

The emergence of composites that exhibit superior mode I fracture toughness relative to first generation brittle thermosets has shifted emphasis to understanding the relationship between the mode II interlaminar fracture toughness and damage tolerance of composite structures. The compression after impact (CAI) test [48,49] has been routinely used to characterize the damage tolerance of composite materials. In this test method, a quasi-isotropic panel is subjected to lateral impact and then loaded in compression to measure residual properties. Low-velocity impact can result in a complex distribution of interior damage that consists of intralaminar cracks, fiber fractures, and multiple delaminations that significantly reduce the residual properties of the laminated composite structures. The residual compressive strength has exhibited good correlation with the mode II interlaminar fracture toughness [50].

The following discussions will focus on mode II interlaminar fracture characterization of composites using the End Notch Flexure (ENF) specimen. This subsection is organized into six segments: Background, Test Method, Deformation Mechanisms, Beam Theory Analysis, Experimental Approaches, and Testing Issues.

Background

From a historical viewpoint, the Short Beam Shear (SBS) test method (ASTM D-2344) was perhaps the first test method used to investigate interlaminar shear fracture of composite laminates on a routine basis [51,52]. Introduced in the 1960s, the method still remains an important quality control test. However, it has some limitations. The small span-to-thickness ratio, in conjunction with perturbations in the stress distribution due to load introduction that do not decay rapidly in orthotropic materials, violates the beam-theory-based data reduction scheme and promotes alternate failure modes [52,53]. Consequently, the SBS test method measures the apparent interlaminar shear strength of the composite.

Classical linear elastic fracture mechanics (see subsection 6.3.1) has been used more recently to characterize the interlaminar fracture toughness of composite materials. In contrast to the SBS test method, well-defined delaminations are embedded or machined into the test specimen. Barrett and Foschi [54] utilized the ENF specimen to characterize the mode II interlaminar fracture of cracked wood beams in 1977. In 1982, Russell and Street [55] used the specimen to characterize mode II critical strain energy release rates of advanced composites. The geometry of the ENF speci-

men (see Figure 6.3-18) is essentially a 3-point flexure specimen with an imbedded through-width delamination placed at the laminate mid-surface. The delamination is placed at the end of the specimen to accommodate the sliding deformation of the sublaminates that results from the flexural loading.

It should be noted that alternate mode II interlaminar fracture flexural specimens have proliferated since the introduction of the ENF specimen to composites. In 1986 Bradley [56] introduced the so-called end-loaded split (ELS) laminate. Prel et al. [57] and Benzeggagh et al. [58] used the ELS specimen and introduced the cantilever beam enclosed notch (CBEN) specimen in 1985. Vu-Khanh [59] also used the CBEN-type specimen in 1987 when he conducted a mode II crack-arrest study. Maikuma et al. [60,61] introduced the Center Notch Flexural (CNF) specimen in 1988 to study mode II interlaminar fracture under static and impact loading. Crews and Reeder [62] introduced the Mixed Mode Bending (MMB) specimen in 1988, which can be used to conveniently measure delamination toughness over a wide range of mode I/II ratios, as well as pure mode I and mode II fracture toughnesses. The flexural specimens are all similar in geometry, as shown in Figure 6.3-19.

Although each specimen possesses certain advantages, the ENF has emerged as one of the most convenient mode II interlaminar fracture specimens. It is easy to manufacture, the test fixturing is simple, and the data reduction methodology is straightforward. For these reasons, the ENF specimen is being included in an international round robin activity with the European Group on Fracture, the Japanese Industrial Standards Group, and ASTM prior to establishing a standard for mode II fracture testing [29]. The MMB specimen is also being included in the round robin test program.

Carlsson and Gillespie [63] have compiled a comprehensive review of fracture mechanics approaches to characterize mode II interlaminar fracture of composites. The review focuses on the ENF specimen. Analytical approaches include shear deformation beam theory; shear deformation plate theory, where the crack tip singularity is introduced as a surface traction; and a higher order beam theory. Numerical results investigating compliance, strain energy release rates, delamination offset from the mid-plane, and frictional effects are summarized. Friedrich et al. [64] focus on the material science issues of mode II interlaminar fracture of PEEK composites. In the following segments, subject

VARIOUS MODE II FLEXURE SPECIMEN

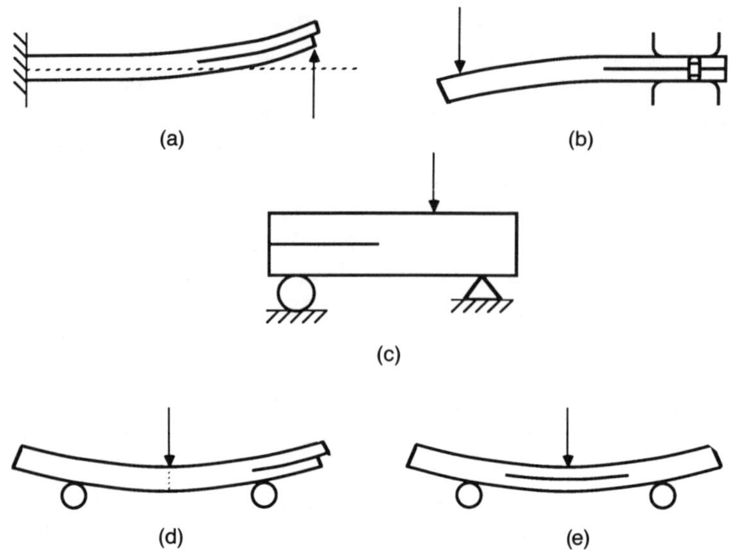

FIGURE 6.3-19. Mode II interlaminar fracture flexural specimens; (a) Bradley [56]; (b) Prel et al. [57,58] and Vu-Khanh [59]; (c) Barrett and Foschi [54]; (d) Russell and Street [55]; (e) Maikuma et al. [60].

matter relevant to the experimental characterization of mode II interlaminar fracture toughness using the ENF test method is emphasized.

Test Method

The geometry of the ENF specimen (see Figure 6.3-18) is essentially a 3-point flexure specimen with an imbedded through-width delamination placed at the laminate mid-surface. The delamination is placed at the end of the specimen to accommodate the sliding deformation of the sublaminates, which results from the flexural loading. A typical ENF specimen is 25 mm wide (w), 125 mm long, and 3–4 mm thick ($2h$). To provide the starter crack, thin inert films, e.g., PTFE and polyimide, should be coated with a release agent and placed at the mid-plane of the laminate. These films are typically 15–75 μm in thickness. The typical ENF test setup is shown in Figure 6.3-20. The fracture test uses the conventional 3-point flexural test fixture with a typical total span, $2L$, of 50.8 mm between the outer specimen support pins. The crack length, a, is 25.4 mm ($a/L = 0.5$) so that the crack tip is located midway between the outer support pin and the midspan load nose.

Specimens should be conditioned to contain a uniform moisture content by drying for 10 days at 77°C. Following the drying cycle, specimens should be stored in a dessicator for up to one day and then removed immediately before testing. Measure the width and thickness of each specimen to the nearest 0.05 mm at the mid-point and at 25 mm from either end. The variation in thickness along the length of the specimen should not exceed 0.1 mm. Average values of the width and thickness measurements should be recorded. Apply a thin coating of brittle typewriter correction fluid, or equivalent, and mark the crack tip location for visual crack determinations. Note that the crack length is defined as the distance from the crack tip to the center of the outer support pin. Consequently, the contact point between the load pin should also be marked on the side of the specimen prior to testing.

The traveling microscope shown in Figure 6.3-20 has proven effective in placing the specimen in the fixture with the desired crack length. The microscope focused on the crack tip also allows crack growth to be monitored during the test. The specimen span-to-depth ratio should be based on the specimen design considerations discussed previously. Cross-head displacement rates should be consistent with the ASTM D-790 recommendations. Prior to testing, the specimen should be pre-cracked as discussed earlier. A real-time analog display of the load versus mid-span deflection should be generated during the test on an *x-y* recorder. Displacements can be determined from an extensometer or LVDT (linear variable differential transformer) attached to the specimen or from the cross-head displacement that has been corrected for machine compliance.

Figure 6.3-21 shows a typical load-deflection curve for an ENF specimen loaded to fracture. As discussed below, unstable delamination growth occurs until the crack is arrested at mid-span due to the compressive load. At this point, the specimen is unloaded as shown in Figure 6.3-21. The critical load at the onset of crack propagation is employed in the data reduction. In ductile systems such as graphite/PEEK, nonlinearities in the load-deflection curve have been observed prior to unstable fracture, where inelastic material response and stable subcritical crack growth may occur before unstable delamination growth [40,65]. After the test has been completed, the specimen should be completely fractured and the actual crack length measured as the average across the width of the sample. Data from spec-

FIGURE 6.3-20. ENF test setup.

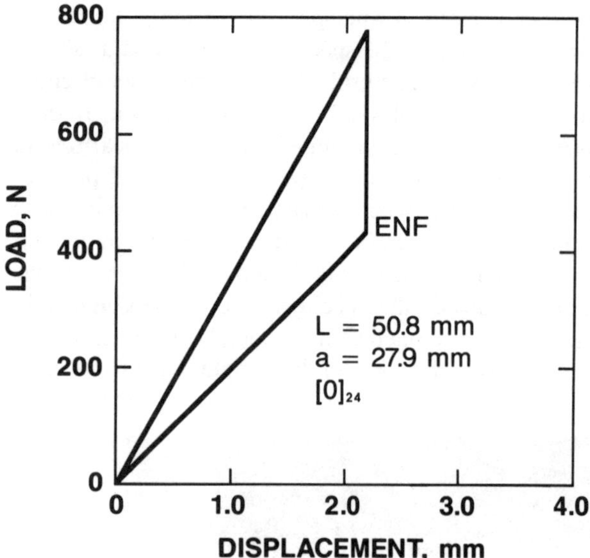

FIGURE 6.3-21. Typical load deflection response of an ENF specimen.

imens exhibiting large variations in crack length across the width should be discarded. Recommended data reduction procedures are summarized in a later discussion.

Deformation Mechanisms

The ENF specimen shown in Figure 6.3-18 is subjected to transverse shear and flexural loadings. Each load type corresponds to an important mechanism that generates crack driving force at the crack tip. Figure 6.3-22 shows the delaminated region of the specimen. The presence of transverse shear forces generates interlaminar shear stresses that distort the beam cross section. This distortion is illustrated in Figure 6.3-22 and contributes to the crack driving force. The dominant mechanism, however, corresponds to the relative sliding deformation (forward shear) between the delaminated sublaminates induced by the lateral load applied at mid-span (see Figure 6.3-22). Alternatively, note that the flexure stresses at the interface are tensile in the upper beam and compressive in the lower beam. At the crack tip, this discontinuity in stress is eliminated by the presence of the interlaminar shear stress singularity associated with mode II crack propagation.

Based upon the physical insight into the sources of the crack driving force, the dominant mechanism based on strength of materials concepts will clearly depend on specimen geometry and material properties. The flexural mechanism, sensitive to parameters that increase the bending deformation, is strongly related to crack length, flexural modulus, and specimen thickness. The transfer shear force mechanism is strongly dependent on the ratio of specimen thickness to crack length and the interlaminar shear modulus of the material.

The mechanics of the ENF geometry are quite complex because of the presence of a crack in a finite domain. Moreover, the beams in the delaminated region, AB in Figure 6.3-18, are elastically built in at the cross

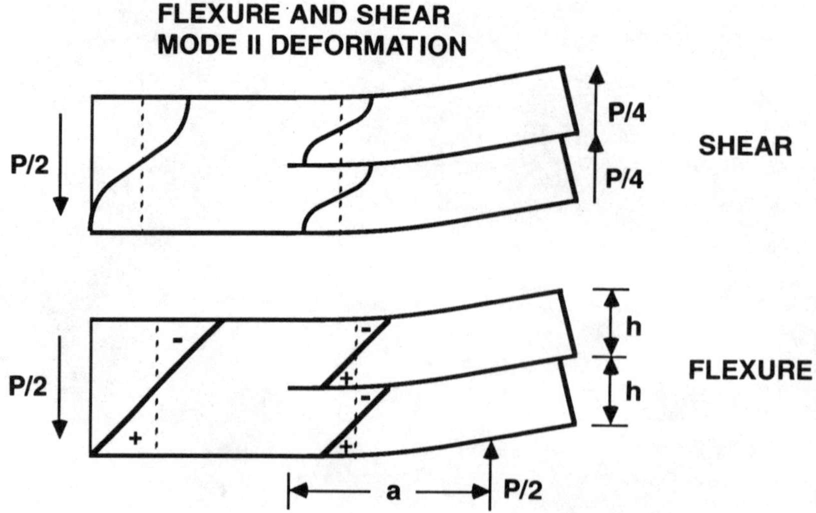

FIGURE 6.3-22. Mechanisms of mode II crack driving force.

section adjacent to the crack tip, which further complicates the analysis. Analytical expressions for compliance and strain energy release rate based upon beam theory (BT), a higher order beam theory (HOBT) based upon Reissner's variational principle, and a shear deformation plate theory (PT) are presented in reference [63]. Beam theory analysis is reviewed next because it is the foundation for the experimental data reduction procedures presented later.

Beam Theory Analysis

The beam theory analysis summarized in this discussion is based upon the model of Carlsson et al. [66]. The model is based upon a shear deformation theory in the uncracked region of the specimen and an approximate elasticity solution for bending of an end-loaded cantilever beam [67].

Friction may dissipate energy in addition to creating a new crack surface. As shown in Figure 6.3-18, the load must be transferred from the lower beam to the upper beam of the delaminated region at the right support (A). The flexure of the beam causes sliding of the crack surfaces, and significant friction between the crack surfaces may dissipate energy, which needs to be accounted for in the analysis. The load transfer between the beams is a problem of significant complexity. A simplified approach will be pursued in the beam theory analysis presented in this subsection. The stability of crack growth and specimen design considerations based upon the model are also discussed.

Compliance. An analytical expression for the compliance of the ENF specimen is required to determine the strain energy release rate by differentiation of this expression with respect to the crack length. Initial beam theory analysis of the ENF geometry [55] did not incorporate shear deformation in the formulation. Incorporation of transverse shear deformation provides a more realistic model for advanced composites, which typically have low interlaminar shear stiffness. The model of Carlsson et al. [66] will be reviewed here. The compliance of the ENF specimen is defined as the displacement, δ, at the central loading pin divided by the applied load, P. Based on the notations defined in Figure 6.3-23, δ may be expressed as

$$\delta = (\Delta_{AB} + \Delta_{BC} + \Delta_{CD})/2 \qquad (6.3\text{-}53)$$

The beams BC and CD are modeled as cantilever

beams based on the assumption that the cross section at C does not warp because it is an approximate line of symmetry. Each beam has an elastic modulus E_1, an interlaminar shear modulus G_{13}, a thickness $2h$, and a width w. Expressions for Δ_{BC} and Δ_{CD} may be obtained from Timoshenko beam theory [68].

For the delaminated region, AB, the displacement Δ_{AB} has two components, i.e., one due to the bending and shearing deformations of the beams and the other due to the rigid body rotation of the cross section at point B (see Figure 6.3-23). In this analysis, the ends of the parallel beams (B in Figure 6.3-18) are assumed to be allowed to deform freely under the action of shearing stress [67]. This assumption is introduced to better model elastic restraint effects at the built-in ends at B, although it is recognized that it may satisfy only pointwise continuity in displacement at the cross section.

Expressions for the displacement contributions in Equation (6.3-53) are provided in reference [66]. Based upon the small deflection assumption, all sources of displacement may be substituted into Equation (6.3-53) to yield the compliance, $C_{SH} = \delta/P$, of the ENF specimen with shear deformation included:

$$C_{SH} = \frac{2L^3 + 3a^3}{8E_1wh^3}\left[1 + \frac{2(1.2L + 0.9a)h^2E_1}{(2L^3 + 3a^3)G_{13}}\right]$$

$$(6.3\text{-}54)$$

For materials with a large shear rigidity and small thickness-to-crack length ratio (h/a), Equation (6.3-54) reduces to the classical beam theory (BT) expression

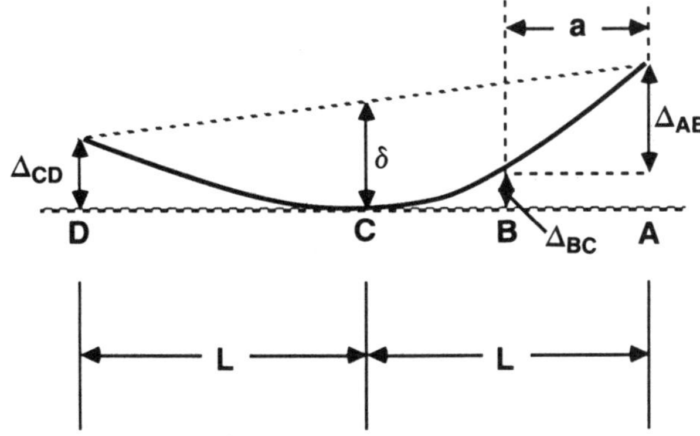

FIGURE 6.3-23. Nomenclature for the ENF: shear deformation beam theory.

for the compliance, C_{BT}, given by Russell and Street [55]:

$$C_{BT} = \frac{2L^3 + 3a^3}{8E_1wh^3} \qquad (6.3\text{-}55)$$

where L is the half-span length (see Figure 6.3-18).

Strain Energy Release Rate. Once the compliance expression has been obtained, the strain energy release rate, G, is evaluated by differentiation of the compliance expression with respect to crack length [see Equation (6.3-25)]:

$$G_{II}^{SH} = \frac{9a^2P^2}{16E_1w^2h^3}\,[1 + 0.2(E_1/G_{13})(h/a)^2] \qquad (6.3\text{-}56)$$

Subscript *"II"* denotes that the ENF test is a pure mode II test. This was verified by finite element analysis of the ENF geometry [69] using the crack closure method [16]. Superscript *"SH"* denotes that shear deformation is included in the energy release rate.

For materials and geometries that result in small values of the quantity $(E_1/G_{13})(h/a)^2$, Equation (6.3-56) reduces to the beam theory, *"BT,"* expression for G_{II} given by Russell and Street [55]:

$$G_{II}^{BT} = \frac{9a^2P^2}{16E_1w^2h^3} \qquad (6.3\text{-}57)$$

Superscript *"BT"* here denotes classical beam theory.

Equations (6.3-55) and (6.3-57), based on classical beam theory, are convenient to nondimensionalize the expressions for compliance [Equation (6.3-54)] and strain energy release rate [Equation (6.3-56)], enabling shear deformation to be quantified. In this manner,

$$\frac{C_{SH}}{C_{BT}} = 1 + \frac{2(1.2L + 0.9a)h^2}{2L^3 + 3a^3}\left(\frac{E_1}{G_{13}}\right) \qquad (6.3\text{-}58)$$

$$\frac{G_{II}^{SH}}{G_{II}^{BT}} = 1 + 0.2\left(\frac{E_1}{G_{13}}\right)\left(\frac{h}{a}\right)^2 \qquad (6.3\text{-}59)$$

Table 6.3-1 illustrates the influence of interlaminar shear on compliance and strain energy release rate for typical ENF geometries ($a/L = 0.5$). For unidirectional specimens, the interlaminar shear modulus can be approximated by the in-plane shear modulus without introducing significant error in the shear correction fac-

tor. The error induced by neglecting shear deformation may be substantial for thick beams of low shear rigidity. The influence of shear deformation becomes increasingly significant for shorter crack lengths [see Equation (6.3-59)]. The shear deformation beam theory expressions for compliance and strain energy release rate in Equations (6.3-54) and (6.3-56), respectively, are used for experimental data reduction purposes.

Stability Considerations for Crack Growth. The crack growth becomes unstable when the net energy supplied exceeds the required crack tip energy [42]. In terms of the Griffith criterion, this may be expressed as

$$\frac{dG_{II}}{da} > 0 \qquad (6.3\text{-}60)$$

provided that G_{IIc} is a constant, independent of crack extension. In reference [66] it was shown that fixed load conditions during ENF testing always yield unstable crack growth. Stable crack growth may be obtained under fixed grip conditions when the crack length to half-span length satisfies the following inequality:

$$\frac{a}{L} \geq \frac{1}{\sqrt[3]{3}} \approx 0.69 \qquad (6.3\text{-}61)$$

ENF testing is not normally conducted in this range because of difficulties in detecting the onset of crack growth and the potential for load introduction effects at mid-span to influence fracture results. For the commonly used test specimen geometry, $a/L = 0.5$, the crack growth is unstable. This conclusion is in agreement with experiments [40].

Influence of Friction. In ENF testing, friction between the crack surfaces in the load transfer region ("A" in Figure 6.3-18) between the delaminated beams may exist. Friction opposes the relative sliding deformations of the beams in the delaminated region and is an energy dissipating mechanism in addition to the energy absorbed in creating new crack surfaces. Friction should not be ignored if accurate fracture data are to be measured.

Static friction corresponds to sticking of the sublaminates and has not been observed experimentally during cyclic loading and unloading of the ENF specimen. Sliding friction, however, would reduce the specimen compliance and might be difficult to observe experimentally. To investigate the influence of sliding friction, the beam theory analysis presented in this subsection is modified to include Coulomb's law for sliding friction:

$$F_f(x) = \mu N(x) \tag{6.3-62}$$

where μ is the coefficient of friction, $N(x)$ is the distribution of normal surface tractions, and x is a coordinate direction defined in Figure 6.3-24. The distribution of normal surface tractions, $N(x)$, is extremely difficult to calculate because of the localized loading and the anisotropic material properties. The numerical finite element analysis of the ENF specimen reported by Gillespie et al. in reference [69] shows that the contact area is less than $4h$ in length where h is the thickness of the beams in the delaminated region. With this insight, the distribution of normal surface tractions was approximated [66] by Dirac's delta function, $\delta(x)$, [70]:

$$N(x) = (P/4)\delta(x) \tag{6.3-63}$$

Integrating the frictional force times the relative sliding deformations over the contact region yields an upper bound estimate of the frictional work because the distributed force is replaced with a point load at the position where maximum sliding deformation occurs [66]:

$$W_f \leq \frac{\mu P \Delta u(0)}{4} \tag{6.3-64}$$

In Equation (6.3-64), $\Delta u(0)$ is the relative sliding deformation of the crack surfaces at the support ($x = 0$ in Figure 6.3-24). The relative sliding deformation was calculated in reference [66] from the cantilever beam analysis given in reference [67]. The result is

$$W_f \leq \frac{3P^2\mu}{8E_1 wh^2}\left[a^2 + \frac{h^2}{12}\left(\frac{E_1}{G_{13}}\right)\right] \tag{6.3-65}$$

This energy was incorporated into a Griffith energy balance in reference [66], giving strain energy available for creating crack surfaces according to

$$\frac{G_{II}^{SH}(\mu)}{G_{II}^{BT}} = 1 + 0.2(E_1/G_{13})(h/a)^2 - 4\mu(h/a)/3 \tag{6.3-66}$$

where $G_{II}^{SH}(\mu)$ denotes strain energy release rate incorporating shear deformation and friction, and G_{II}^{BT} is the

Table 6.3-1. Influence of interlaminar shear deformation on the compliance and the strain energy release rate of the ENF specimen [66].

L, mm	a, mm	h, mm	E_1/G_{13}	C_{SH}/C_{BT}	G_{II}^{SH}/G_{II}^{BT}
50.8	25.4	1.52	18.3	1.023	1.013
50.8	25.4	1.52	26.9	1.034	1.019
38.1	19.3	1.52	18.3	1.041	1.023
38.1	19.3	1.52	26.9	1.060	1.033
38.1	19.3	2.61	12.8	1.084	1.047
38.1	19.3	2.61	25.7	1.168	1.094

beam theory expression defined in Equation (6.3-57).

The influence of friction is best illustrated by defining a nondimensional energy release rate parameter, $g(\mu)$:

$$g(\mu) = [G_{II}^{SH} - G_{II}^{SH}(\mu)]/G_{II}^{BT} \tag{6.3-67}$$

Substitution of Equations (6.3-56) and (6.3-57) into Equation (6.3-67) yields

$$g(\mu) = 4\mu(h/a)/3 \tag{6.3-68}$$

This equation shows that the reduction in available strain energy is directly proportional to μ and decreases

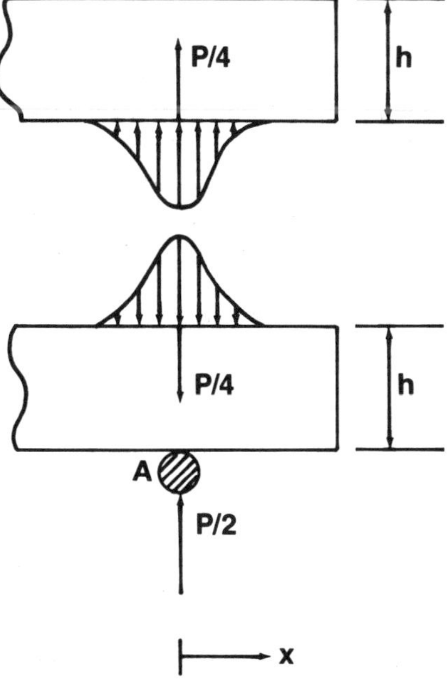

FIGURE 6.3-24. Normal stress distribution between delaminated beams.

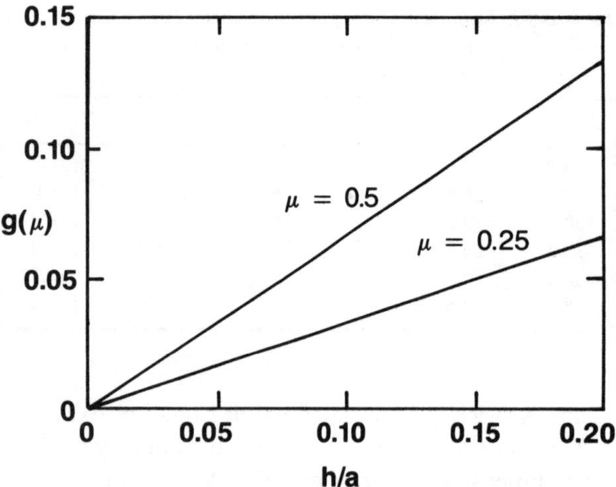

FIGURE 6.3-25. Influence of friction between crack surfaces: shear deformation beam theory.

with increased crack length-to-thickness ratio (a/h). $g(\mu)$ is plotted versus h/a in Figure 6.3-25 for two reasonable coefficients of sliding friction, $\mu = 0.25$ and 0.5. Typical ENF geometries have h/a less than 0.05 [42], and, for those geometries, the error induced in G_{II}^{SH} is only 2 to 5 percent according to this analysis. Experimental confirmation of these results has been reported by Russell and Street [71], where hysteresis in the loading and unloading curves indicates less than a 2 percent overestimate of G_{II}^{SH}.

Specimen Design Considerations. All analysis results presented up to this point are based on small strain and small deflection theory and assumption of linear elastic material response up to fracture. Geometrical non-linearities due to large displacements and rotations complicate the analysis significantly and can be avoided by using thick beams or short spans [66]. It is also important that interlaminar fracture occur prior to flexural failure of the beam. Significant nonlinear material response due to yielding of the material in tension or compression must also be minimized [66]. As will be shown, the requirement of linear elastic behavior leads to lower bounds on beam thickness. In reference [72], bounds on crack length were derived based on minimization of friction and interactions between the crack tip singular stress and concentrated load effects due to the center load nose so that the assumptions in the beam theory formulation are not violated. These specimen design considerations will be reviewed in this discussion.

Sizing the Thickness. In beam theory the exact expression for the curvature $\varkappa$ is

$$\varkappa = \frac{1}{\varrho} = \frac{v''}{[1 + (v')^2]^{3/2}} \tag{6.3-69}$$

where $v = v(x)$ is the deflection curve, ϱ is the radius of curvature, and prime and double prime denote differentiation once and twice, respectively, with respect to x, a coordinate direction along the beam axis.

For small deflections/rotations, the slope v' is small so that $(v') \ll 1$. In Equation (6.3-69) the quantity $(v')^2$ can be neglected:

$$\varkappa = v'' \tag{6.3-70}$$

This constitutes the classical approximation in linear small deflection beam theory. In reference [66] the minimum semi-thickness, h_{min}, of the beam was calculated by classical beam theory as a function of the maximum allowable slope of the deflection curve, v'_a, the fracture toughness, G_{IIc}, to be measured, and specimen material and geometrical parameters:

$$h_{min} = \left[\frac{G_{IIc}(L^2 + 3a^2)^2}{4(v'_a)^2 a^2 E_1} \right]^{1/3} \tag{6.3-71}$$

A second thickness requirement may also be formulated based on the requirement that the maximum bending strain be less than the proportional limit strain or the ultimate strain of the material. Calculations presented in reference [66] yield

$$h_{min} = \frac{L^2 G_{IIc}}{a^2 \epsilon_{m,a}^2 E_1} \tag{6.3-72}$$

where $\epsilon_{m,a}$ is defined as the strain allowable required to maintain linear elastic material response. Equations (6.3-71) and (6.3-72) are presented in Figure 6.3-26, where h_{min} is plotted versus G_{IIC} for two typical span lengths ($L = 38.1$ mm and 50.8 mm, $a/L = 0.5$, $v_a = 0.2$, and $e_{m,a} = 1\%$). Results presented in Figure 6.3-26 correspond to a ply thickness of 0.127 mm. The error in the curvature expression is less than 6% when the maximum allowable slope, v'_a, is ≤ 0.2. G_{IIC} for graphite fiber reinforced thermoset and thermoplastic composites is commonly less than 4 kJ/m² [2, 40]. In this range of G_{IIC}, the small deflection requirement, Equation (6.3-71), dictates the minimum number of plies required in the ENF specimen.

Admissible Crack Lengths. Constraints on crack length were analyzed in reference [72]. A lower bound on crack length is dictated by the influence of friction discussed previously. Here the strain energy release rate parameter $g(\mu)$ defined in Equation (6.3-68) is the starting point. As stated previously, friction is an energy absorbing mechanism that would increase the apparent fracture toughness if not accounted for properly in the data reduction procedure. For accurate and straightforward data reduction, frictional effects should be minimized by appropriate specimen design. Given that friction between the crack surfaces should not reduce the available energy release rate by more than a small fraction, f, Equation (6.3-68) yields

$$a_{min} = \frac{4\mu h}{3f} \qquad (6.3\text{-}73)$$

Note that the minimum crack length is directly proportional to the specimen semi-thickness.

An upper bound on crack length is, by similar reasoning, given by the condition that the sliding crack surfaces not be inhibited by the compressive stress field at the center load nose. This was investigated by Gillespie et al. [72], who found that the total contact length is less than $4h$, consistent with the results presented in reference [69]. Since the contact pressure distribution is symmetric around the point of load introduction at mid-span, a maximum crack length, a_{max}, is given by

$$a_{max} = L - 2h \qquad (6.3\text{-}74)$$

Equations (6.3-73) and (6.3-74) provide useful specimen design information for allowable crack lengths. For a typical twenty-four-ply graphite/epoxy ENF geometry ($h = 1.7$ mm and $L = 50.8$ mm), Equations (6.3-73) and (6.3-74) ($\mu = 0.25$ and $f = 3.3\%$) identify the following admissible range of crack lengths:

$$17 \text{ mm} \leq a \leq 47.4 \text{ mm} \qquad (6.3\text{-}75)$$

Note that $a/L = 0.5$ falls within these bounds in this example.

It should be recognized that crack lengths larger than $a/L \cong 0.69$ result in stable crack growth. This may be undesirable from a data reduction point of view as it becomes increasingly difficult to define the critical load

[40]. From an experimental viewpoint, this may represent a stringent constraint on the maximum admissible crack length. For the previous example, a_{max} is reduced to approximately 35 mm.

As discussed in references [63,72], the applicability of the plate theory singularity approach potentially imposes further constraints on admissible crack lengths. Plate theory assumes that the stress singularity due to the crack tip decays to a constant shear stress given by classical beam theory. Obviously, if the crack is long, the crack tip stress field will interact with the contact stress field due to the center point load as shown in reference [73], and the assumption of a stress decay to a value given by classical beam theory will be in error. This could be avoided by imposing restrictions on the maximum allowable crack length as discussed in references [63,72]. The influence of stress decay and inelastic materials behavior on admissible crack lengths is discussed next.

Summary. The analysis presented in this discussion suggests that interlaminar shear deformation may influence the compliance and strain energy release rate of the ENF specimen. Neglecting interlaminar shear deformation leads to an underestimate of the compliance and strain energy release rate that may be significant for advanced composite materials exhibiting a high degree of anisotropy. For commonly used unidirectional lay-

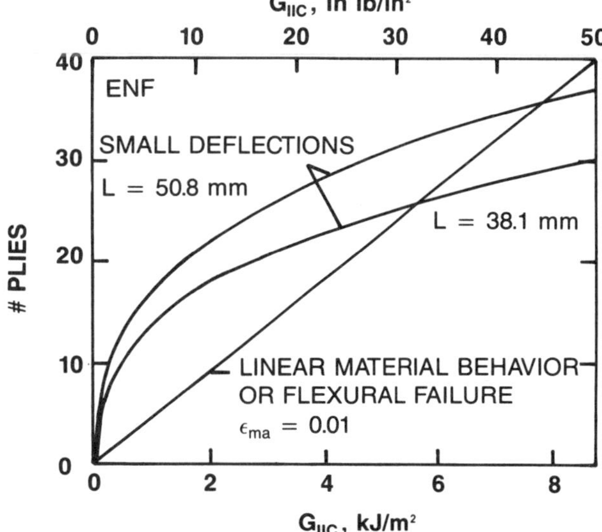

FIGURE 6.3-26. ENF specimen thickness must be chosen to minimize geometric nonlinearities and maintain linear material response.

Table 6.3-2. Correlation of finite element results [74] to shear deformation beam theory [66].

a/L	Glass/Epoxy		Graphite/Epoxy	
	G_{II}^{FE}/G_{II}^{BT}	$(G_{II}^{FE} - G_{II}^{SH})/G_{II}^{FE}$ (%)	G_{II}^{FE}/G_{II}^{BT}	$(G_{II}^{FE} - G_{II}^{SH})/G_{II}^{FE}$ (%)
0.2	1.17	5.43	1.34	1.64
0.4	1.07	3.99	1.16	6.67
0.6	1.04	2.41	1.10	5.76
0.8	1.02	1.39	1.07	4.54
0.9	1.01	0.91	1.05	3.40

ups and geometries [66], however, the error in toughness value induced by neglecting shear deformation is less than 10 percent according to this analysis. Errors in compliance are approximately twice the errors in toughness for the geometries in Table 6.3-1. Shear deformation should most certainly be included in the experimental approaches discussed previously for characterizing the mode II interlaminar fracture toughness.

A cursory examination of Equation (6.3-59) for the strain energy release rate indicates that the thickness-to-crack length ratio (h/a) also significantly influences the magnitude of the strain energy release rate relative to classical beam theory. In the limit, as a/h approaches zero, the shear correction increases dramatically. In Table 6.3-2, the finite element results of Salpekar et al. [74] are compared to the shear deformation beam theory of Carlsson et al. [66] over a range of crack lengths (0.2 ≤ a/L ≤ 0.9). Reasonable agreement is obtained for typical glass/epoxy and graphite/PEEK composites, lending confidence to the accuracy of the analytical solution over a wide range of crack lengths and material systems.

Specimen design considerations include geometric nonlinearities and friction between the crack surfaces. Geometric nonlinearities may be important when characterizing tough material systems. Friction is important to consider when testing thick beams and short crack lengths. The procedures presented in this discussion should be used to select specimen thicknesses and crack lengths to maintain linear elastic small deflection response and to minimize errors due to friction.

Experimental Data Reduction Procedures

In this segment, experimental data reduction methodologies and testing issues are presented for characterizing mode II interlaminar fracture of composites using the ENF specimen. The data reduction schemes are based upon either the analytical beam theory expressions presented previously or traditional compliance calibration techniques. Modifications to the data reduction procedures to account for nonlinear load-deformation response are also presented.

Beam Theory. The starting point for the data reduction scheme is beam theory including shear deformation, reviewed earlier in this discussion. From the slope of the load (P) versus deflection (δ) curve, the compliance, C, may be determined experimentally as $C = \delta/P$. Assuming $C = C_{SH}$ yields

$$C = C_{SH} = \frac{2L^3 + 3a^3}{8E_1wh^3} + \frac{1.2L + 0.9a}{4whG_{13}} \quad (6.3\text{-}76)$$

One can then solve for the flexural modulus required in the expression for strain energy release rate. Substitution into Equation (6.3-56) for G_{II}^{SH} yields

$$G_{II}^{SH} = \frac{9a^2P^2}{4wL^3} \frac{(C - C_{SH}^*)}{\left[1 + 1.5\left(\frac{a}{L}\right)^3\right]} \quad (6.3\text{-}77)$$

C_{SH}^* is a modified shear compliance given by

$$C_{SH}^* = \frac{6L + 3a - \dfrac{L^3}{a^2}}{20whG_{13}} \quad (6.3\text{-}78)$$

For unidirectional specimens the interlaminar shear modulus, G_{13}, can be approximated by the in-plane shear modulus without introducing significant error in the shear correction C_{SH}^*. From an experimental viewpoint, the compliance expression must also be corrected for machine compliance, C_m. The major advantage to this approach is that only a single test is required to characterize the critical strain energy release rate.

Compliance Calibration. Compliance calibration represents an alternative way to reduce fracture toughness data. In this approach, compliance is measured for several crack lengths and curve fit to a polynomial expression to calculate the coefficients. The resulting compliance expression is then differentiated with respect to crack length to yield dC/da and, hence, G_{IIC} through Equation (6.3-25). This method has been employed successfully in ENF testing [40,65].

Compliance calibration requires a long ENF specimen with a long crack so that various crack lengths can be achieved by sliding the specimen (see Figure 6.3-27). When compliance data are generated, the specimen must not be overloaded or crack extension may occur. This method has the advantage that it does not require an analytical expression for the compliance, since the compliance versus crack length relation is obtained experimentally. Furthermore, errors due to test machine compliance are eliminated in this approach, since the compliance relation is differentiated to achieve G_{IIC}. The disadvantage, however, is that many tests must be conducted to generate the compliance versus crack length data. The flexural test results are also very sensitive to fluctuation in specimen thickness.

The order of the polynomial used to fit C vs. a data is commonly guided by an analytical approach. Based upon classical beam theory, a third order polynomial is sufficient for the ENF specimen [see Equation (6.3-55)]:

$$C = C_1 + ma^3 \qquad (6.3\text{-}79)$$

where C_1 is a constant that includes machine compliance and m is the slope of the line C vs. a^3. The resulting expression for the strain energy release rate based on compliance calibration (*cc*) is

$$G_{II}^{cc} = 1.5P^2ma^2/w \qquad (6.3\text{-}80)$$

An alternative way to reduce the data is to normalize the compliance with the compliance C_0 for the beam with no crack [40]:

$$C/C_0 = 1 + m(a/L)^3 \qquad (6.3\text{-}81)$$

The parameter m is obtained as the slope of the line C/C_0 vs. $(a/L)^3$ (see Figure 6.3-28). As a comparison,

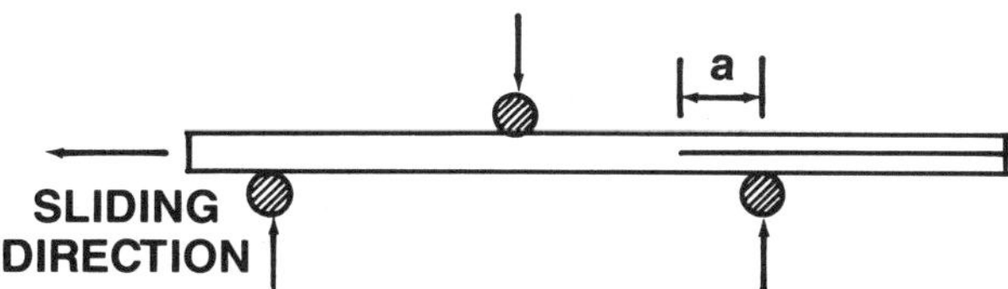

FIGURE 6.3-27. Compliance calibration requires specimen with long crack length.

the classical beam theory expression, Equation (6.3-55), gives a slope $m = 1.5$. Differentiation with respect to crack length in conjunction with Equation (6.3-25) yields

$$G_{II} = \frac{3mP^2a^2C_0}{2wL^3} \qquad (6.3\text{-}82)$$

This method was applied in reference [40] on graphite/epoxy and graphite/PEEK composites. Slopes ranging from 1.42 to 1.51 were obtained for four specimens tested, indicating that classical beam theory is a reasonable approximation to the ENF compliance. A comparison between G_{IIC} reduced by Equation (6.3-77) and Equation (6.3-80) is presented for a graphite/epoxy composite in Table 6.3-3. As observed, there is some scatter between the methods, but no trend can be observed indicating that either method can be used to reduce G_{IIC}.

Alternative curve fitting methods were employed by O'Brien et al. [65]. They performed compliance cali-

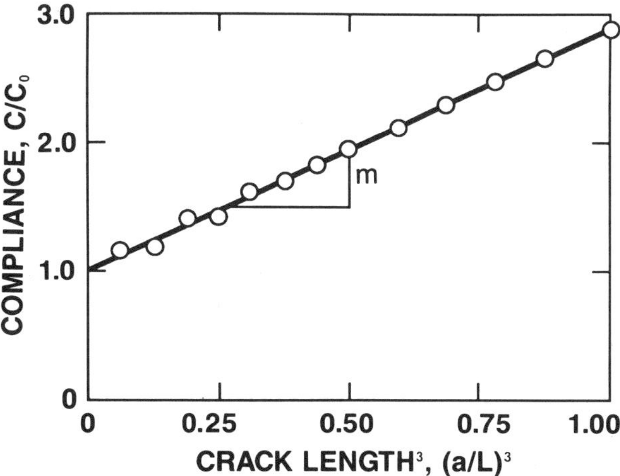

FIGURE 6.3-28. Compliance calibration: C/C_0 versus $(a/L)^3$ for the ENF specimen.

Table 6.3-3. Comparison of fracture toughness in mode II determined from beam theory including shear and compliance calibration. Material is AS4/BP907, after reference [75].

Specimen	SH G_{IIC} kJ/m²	CC G_{IIC} kJ/m²
1	1.033	1.062
2	1.008	1.040
3	1.067	1.054
4	1.089	1.024
5	1.031	0.976

bration based on insight obtained from classical beam theory (BT) and beam theory incorporating shear (SH) according to the following equations:

$$BT:\ C/C_0 = A_0 + A_3(a/L)^3 \qquad (6.3\text{-}83)$$

$$SH:\ C/C_0 = A_0 + A_1(a/L) + A_3(a/L)^3 \qquad (6.3\text{-}84)$$

Here, C_0 represents the compliance measured for a beam with no crack. The form of these equations is easily verified from Equations (6.3-55) and (6.3-76). In the comparison, several material systems and the influence of the support, i.e., roller and knife-edge fixtures, were investigated. Average results for the three material systems studied are given in Tables 6.3-4 and 6.3-5. The values of A_0, A_1, and A_3 listed in Tables 6.3-4 and 6.3-5 were determined from experimental data by least squares regression analysis. Also listed are the theoretical values of A_0, A_1, and A_3 determined from Equations (6.3-55) and (6.3-76) using the appropriate material and geometrical properties [65].

Table 6.3-4. Compliance calibration results based on classical beam theory [Equations (6.3-83) and (6.3-78)].

Material	Roller Fixture		Knife-Edge Fixture		Equation (6.3-55)	
	A_0	A_3	A_0	A_3	A_0	A_1
Glass/Epoxy (S2/SP250)	0.993	1.191	1.000	1.317	1.0	1.5
Graphite/Epoxy (T300/BP907)	1.024	1.305	0.993	1.317	1.0	1.5
Graphite/PEEK (AS4/PEEK)	0.997	1.464	1.006	1.671	1.0	1.5

With respect to classical beam theory, Table 6.3-4 shows that the coefficient A_0 is in close agreement with theory, while A_3 is less than the theoretical value of 1.5. The same data fitted to Equation (6.3-84) gave the results in Table 6.3-5. It should be pointed out that the parameter A_1 representing shear in Equation (6.3-76) cannot be accurately determined because of its small magnitude. Again, the experimentally determined coefficients are in the same order as the theoretical values.

Comparison of Approaches. Various data reduction schemes are compared in Figure 6.3-29. It should be pointed out that all data are determined from specimens with shear driven precracks that produced larger mode II toughnesses than a tension precrack (see references [40,65]). Tough resins tend to produce nonlinearities in the load-displacement record [40,65]. This complicates the data reduction methodology because the linear elastic fracture-based methods discussed herein do not appropriately incorporate change in energy with crack length beyond the onset of nonlinearity (P_{NL}). If it is assumed that the nonlinearity is caused by subcritical crack growth [40,65], a conservative toughness value may be determined by substituting $P = P_{NL}$ and the initial compliance in the linear elastic formulas for G_{IIC}. This is the method used to reduce the data for AS4/PEEK shown in Figure 6.3-29. As an upper bound, G_{IIC}, based on the maximum load and initial compliance that corresponds to unstable delamination growth, P_c, is also shown.

For all three materials, the compliance calibration methods produce lower and, hence, more conservative values than the beam theory values without (BT) and with (SH) shear. Generally, it appears that the most accurate and conservative method to reduce toughness data is to perform compliance calibration and to calculate G_{IIC} using the load at the onset of nonlinearity. However, the gain in accuracy has to be judged against the significantly more extensive test procedure required for the compliance calibration method. The sensitivity of compliance to specimen thickness variations due to volume fraction gradients would require calibration tests on each specimen to obtain accurate results. Furthermore, the ASTM D30.02.02 Task Group on Interlaminar Fracture Toughness [36] has conducted an interlaboratory ENF round robin test program comparing the compliance calibration [Equation (6.3-79)] to the beam-theory-based data reduction including shear deformation [Equation (6.3-77)]. They found that both methods yielded similar mean values and standard deviations. However, the scheme based on Equation

Table 6.3-5. Compliance calibration results based on beam theory incorporating shear deformation [Equation (6.3-78) and (6.3-84)]. Roller fixture only.

Material	Roller Fixture			Beam Theory and Shear		
	A_0	A_1	A_3	A_0	A_1	A_3
Glass/Epoxy (S2/SP250)	1.007	-0.0543	1.239	1.000	0.050	1.399
Graphite/Epoxy (T300/BP907)	0.988	0.1383	1.180	1.000	0.044	1.412
Graphite/PEEK (AS4/PEEK)	1.013	-0.0615	1.519	1.000	0.054	1.408

(6.3-77) is simpler, gives less scatter between laboratories, and is less sensitive to material variability. The task group recommendation is to eliminate compliance calibration in favor of the shear deformation beam theory [see Equation (6.3-77)].

Testing Issues

Interlaminar fracture toughness is extremely sensitive to numerous energy absorbing micromechanisms induced by the applied loads. Friedrich et al. [46,64] considered the micromechanisms shown schematically in Figure 6.3-30, which include (1) formation of the fracture surface of the main crack, (2) plastic deformation of the matrix in the crack tip region, (3) microcracking and secondary cracks, and (4) crack bridging by fibers resulting in peeling and fracture. All of these mechanisms may be present to a certain extent and contribute to a nonlinear load-deformation response during testing.

Toughness is most sensitive to the volume of matrix material undergoing inelastic material deformation. Crack tip plasticity is strongly dependent on the local micro-structure. In general, the size of the plastic zone is inversely proportional to fiber volume fraction due to constraint imposed by the fibers. During processing of fracture specimens, however, the local micro-structure

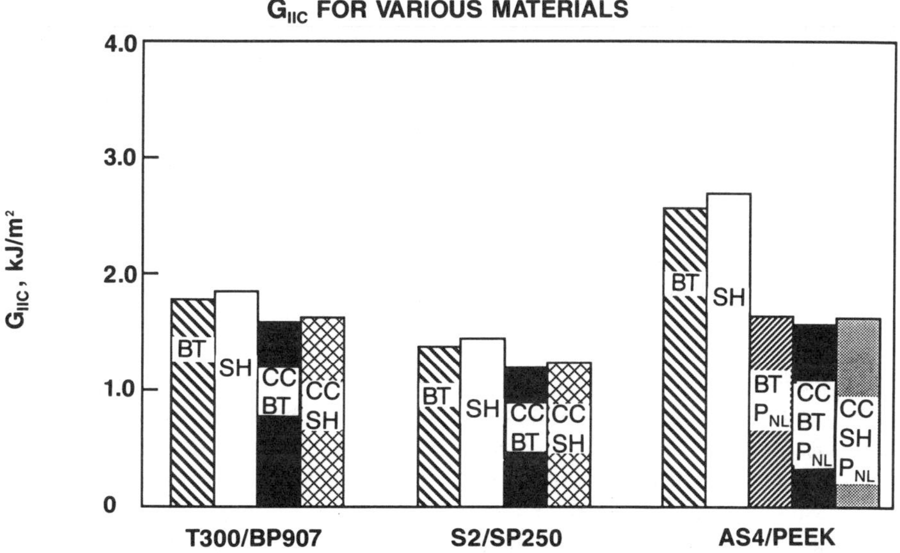

FIGURE 6.3-29. Interlaminar mode II fracture toughness [66] based upon various data reduction schemes. *BT*: Equation (6.3-57), *SH*: Equation (6.3-56), cc(*BT*): Equation (6.3-83), cc(*SH*): Equation (6.3-84).

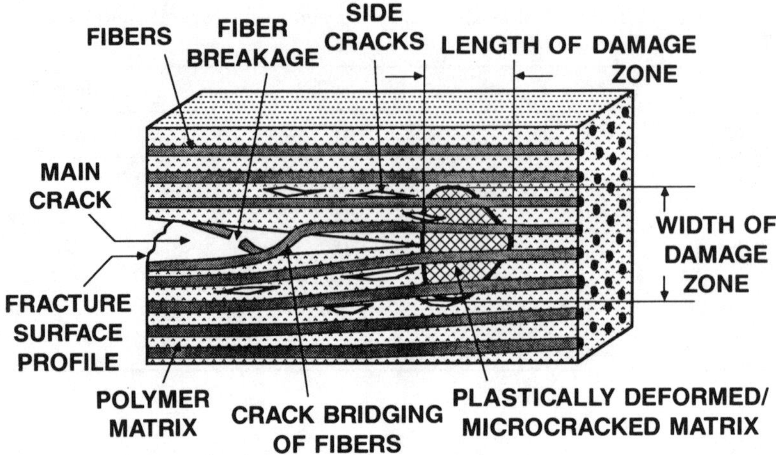

FIGURE 6.3-30. Schematic illustration of energy absorbing mechanisms of interlaminar fracture in unidirectional composites [46].

may be altered by the presence of inserts used for starter cracks during testing. Thick inserts may result in resin rich regions that locally increase composite toughness, since a larger volume of matrix is plastically deformed upon crack extension (see Figure 6.3-8). The volume of material undergoing plastic deformation is also dependent on the amount of crack propagation. Similar to the fracture of ductile metals, crack tip plasticity develops with increasing crack growth. Interlaminar fracture toughness would increase from an initiation value to a propagation level resulting in a classic R-curve response. The material may be viscoelastic and/or viscoplastic, and, therefore, the fracture toughness would be dependent on temperature and deformation rate.

In this discussion, testing issues related to mode II interlaminar crack driving forces are presented. Emphasis is placed on the influence of precracking techniques, characterization of nonlinear load-deformation response, and R-curve effects on mode II interlaminar fracture toughness characterization.

Influence of Precracking Method. One of the most important factors to consider prior to testing an ENF specimen is the method of precracking. For specimens without precracking, G_{IIc} may be artificially higher than that of specimens with natural crack tip morphologies [40]. During processing, a resin rich pocket at the end of the insert is created that blunts the crack tip and increases the resistance to crack growth. The volume of matrix that migrates to the crack tip during processing

is very sensitive to the insert thickness. In reference [75], fracture toughness determined in the ENF test with no precrack was directly proportional to insert thickness.

Consequently, similar to the situation for fracture testing of metals [76], a standard procedure has to be used to precrack the specimen, i.e., extend the starter crack beyond the resin pocket at the end of the insert. In materials that exhibit an increase in fracture toughness with crack extension (R-curve), the crack should be extended as little as possible to ensure accurate characterization of the initiation toughness. An infinitely thin insert would be ideal. In this case, a matrix rich region would not exist, precracking would not be required, and initiation and propagation toughness could be measured. Various types of inserts have been employed including Kapton and aluminum (films and sprays) with thicknesses as small as 12.5 μm.

Several precracking methods have been proposed. The most common method is to clamp the specimen across the width at some short distance ahead of the insert and then wedge the crack surfaces open to propagate a sharp crack a distance between 2 and 5 mm until arrested at the clamp. This approach creates a "mode I" precrack. In materials that are susceptible to fiber bridging, tensile precracking may excite fiber peeling and fracture micromechanisms that contribute to the measured fracture toughness and that are not characteristic of mode II interlaminar fracture in general.

A second method to introduce a precrack is to employ an ENF specimen with a long insert crack. Stable crack growth may be achieved by placing the end of the starter crack close to the center load nose ($a/L > 0.69$). Loading the specimen in 3-point bending slowly propagates the crack from the end of the insert to the center load nose where it is arrested by the compressive stress field. This creates a "mode II" type precrack.

A third way to achieve a well-defined sharp starter crack is identical to the second method except that the precrack is grown by applying a cyclic load [65]. With a high cyclic load level, only a few load cycles are required to propagate the crack from the end of the insert to the center load nose. This creates a "cyclic mode II" precrack.

In materials exhibiting an *R*-curve, such as graphite/PEEK, initiation toughness may be impossible to determine with any of these precracking techniques, since the crack tip plastic zone has evolved to some extent. In this case an infinitely thin insert and no precracking would be desirable. The ASTM D30.02.02 Task Group on Interlaminar Fracture Toughness, in conjunction with the European Group on Fracture and the Japan Industrial Standards Group [29], is currently addressing these issues.

The mode I precrack has the advantage that the crack front can be easily detected on the fracture surfaces after the fracture test is completed. As shown in Figure 6.3-31, mode I and mode II fracture surfaces are easily distinguishable, which makes the determination of initial crack length more accurate. After the fracture test is completed, the initial crack length can be evaluated as the distance between the end of the mode I precrack and the imprint on the specimen surface of the outer support pin. In this way the straightness of the delamination front can be confirmed, and the crack length estimation based on observation of the specimen edges can be verified. In the case of a static mode II precrack, however, the determination of crack length has to rely solely on visual crack tip determination at the edges. The accuracy of the initial crack length determination may be increased by painting the edges with white brittle paint and using a traveling microscope as shown in Figure 6.3-20. The third type of precrack, "cyclic mode II," has the advantage over the static mode II precrack in that the delamination front can be identified on the fracture surface with the same accuracy for initial crack length determination as for the mode I precrack [65]. The disadvantage is that fatigue precracking is more difficult, time-consuming, and costly.

The influence of the type of precrack on the mode II toughness has been investigated in references [40,65]. Table 6.3-6 summarizes G_{IIC} for graphite/PEEK and graphite/epoxy specimens with no precrack and static mode I and mode II precracks. The very large toughness values for the specimens without precracks are evidently due to the blunted crack tip at the end of the insert film. Mode I precracks produced consistent toughness values at different thicknesses and spans but were somewhat lower than the toughness values obtained from specimens with mode II precracks. As pointed out in reference [65], a mode I precrack may not be representative of a naturally occurring delamination in a composite structure because this would require large tensile deformations at the crack tip. A mode II precrack is probably more representative of a delamination in a region of high shear stress. O'Brien et al. [65] investigated the mode II toughness for specimens precracked in cyclic mode II loading. It was found that these specimens had similar toughness values to those precracked under static mode II conditions.

It should finally be pointed out that a correct initial crack length is crucial for the correct determination of toughness. A simple propagation of error analysis of the formulas used to reduce G_{IIC} shows that error in crack length generally dominates the total error in G_{IIC}.

Nonlinear Response. Thermoplastic composites such as graphite/PEEK tend to produce nonlinearities in the load-displacement record during mode II interlaminar fracture testing [40,65] as shown schematically in Figure 6.3-32. This complicates the data reduction methodology summarized previously because the linear elastic fracture-based methods discussed herein do not appropriately incorporate change in energy with crack

Table 6.3-6. Mode II toughness for graphite/epoxy (CCYCOM 982) and graphite/PEEK (APC-2) with various precracks. $a/L = 0.5$. Cross-head rate is 1.25 mm/min [40].

Material	No. of Plies	Precrack	*L* mm	*BT* G_{IIC} kJ/m²
CYCOM 982	24	No	38/51	1.43 ± 0.17*
CYCOM 982	24	Mode I	51	0.68 ± 0.02
CYCOM 982	24	Mode II	38/51	0.73 ± 0.09*
APC-2	26	No	51	2.73 ± 0.33
APC-2	26/40	Mode I	38/51	1.85 ± 0.11**
APC-2	26	Mode II	51	1.93 ± 0.28

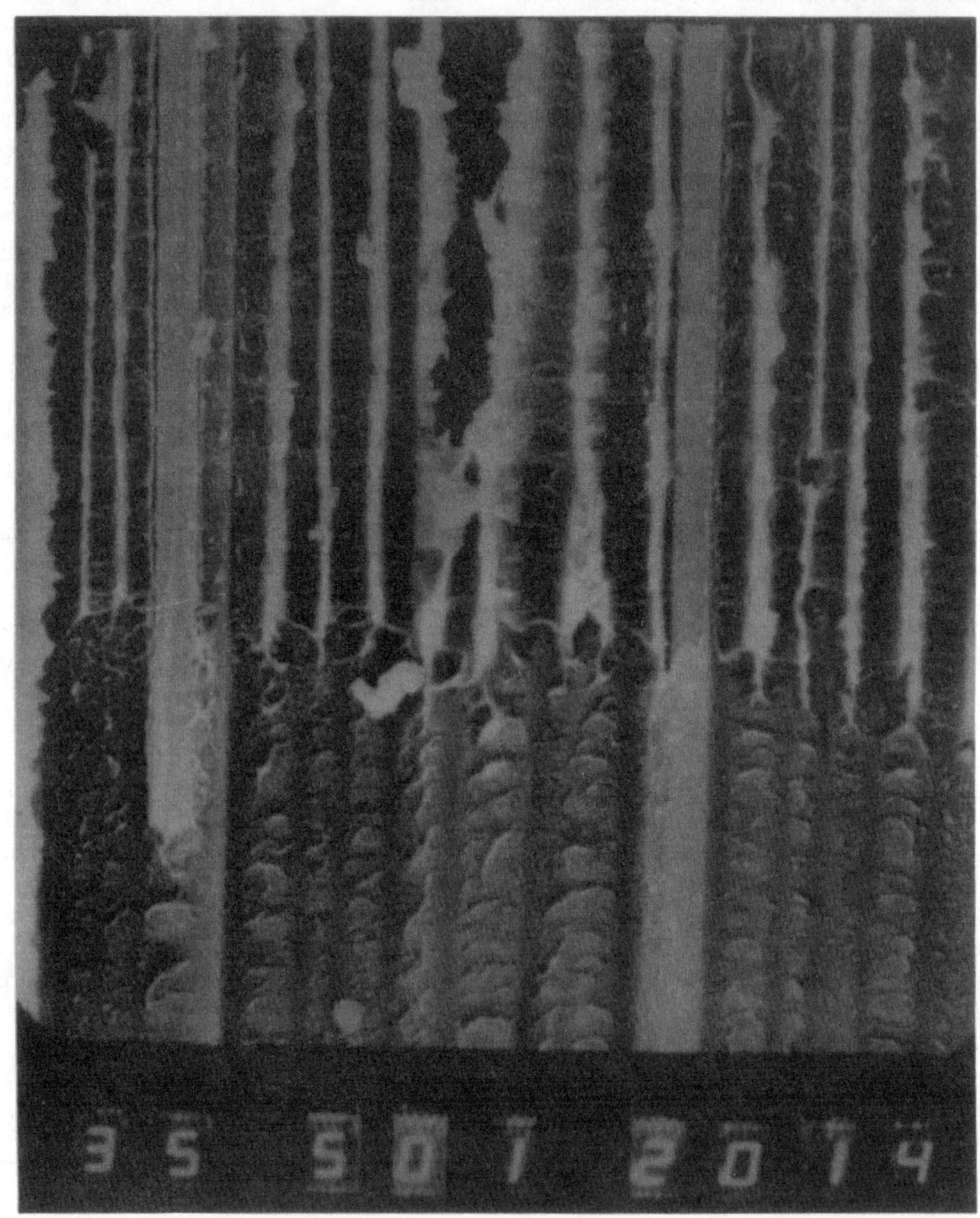

FIGURE 6.3-31. Transition from mode I precrack to mode II interlaminar fracture in graphite/PEEK enables initial crack lengths to be quantified straightforwardly.

length beyond the onset of nonlinearity (P_{NL}). The origin of this nonlinear response can be attributed to numerous mechanisms: (1) material inelastic behavior at the crack tip, (2) subcritical crack growth, (3) rising R-curve, (4) peeling and fracture of fibers (mode I precrack), and (5) nonuniform crack growth across the specimen width. The observed nonlinearities, which are rate dependent [40], are attributed primarily to a combination of slow stable crack growth preceding unstable crack growth and material inelastic behavior (viscoelastic or plastic shear response and damage) in the process zone that evolves around the crack tip.

Nonlinear Fracture Toughness. In thermoplastic composites that exhibit moderate nonlinear response, it is common practice to retain the simplicity of linear elastic fracture mechanics and report an upper bound on G_{IIC} based on the maximum load that corresponds to unstable delamination growth, P_c, shown in Figure 6.3-32. A conservative toughness value designated G_{IISC} may be defined by substituting $P = P_{NL}$ in the linear elastic expression for mode II interlaminar fracture toughness [Equation (6.3-77)]. In both cases, the initial compliance of the specimen is employed.

For materials undergoing substantial nonlinear response, an appropriate definition of fracture toughness is required. Several nonlinear fracture toughness measures, such as the Crack Growth Resistance (R-curve), Crack Opening Displacement (COD), J-integral methods, and Irwin's plastic zone size corrected value of fracture toughness, have been proposed. A relatively recent nonlinear fracture toughness measure, here designated by G_{CNL}, has been examined in some detail by Liebowitz et al. [77,78]. This approach accounts for both crack region inelastic material behavior and subcritical crack growth within the process zone and is based on the entire load-displacement response. This method appears to be appropriate for tough thermoplastic composites. A recent investigation [79] shows that the Liebowitz method yields similar fracture toughness to the J-integral due to similar definitions.

By the Liebowitz method, the deviations from linearity of the load-displacement curve in a fracture test are quantified by a correction factor to the elastic strain energy release rate G_{IIC}. The nonlinear fracture energy, G_{IICNL}, is

$$G_{IICNL} = (1 + \beta)G_{IIC} \qquad (6.3\text{-}85)$$

where β defines a nonlinear correction to the elastic

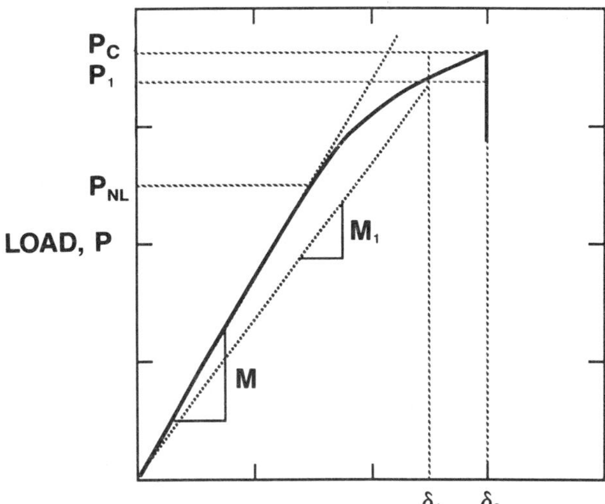

FIGURE 6.3-32. General form of the load-displacement curve for an ENF specimen. M is the initial stiffness, and M_1 is the secant stiffness defined at an arbitrary point within the nonlinear region (P_1,δ_1). P_{NL} is the load at the onset of nonlinear response, and (P_c,δ_c) corresponds to the onset of unstable crack growth.

strain energy release rate. A method to evaluate β for situations where critical crack growth occurs under fixed grip conditions was outlined in reference [77]. Based on this method, β is determined from the load-displacement curve (Figure 6.3-32) in a straightforward manner. The load-displacement (P-δ) curve is assumed to be empirically described by the following three-parameter relation:

$$\delta = P/M + k(P/M)^n \qquad (6.3\text{-}86)$$

where the parameters k and n characterize the deviation from linearity, and M is the initial stiffness of the specimen.

To determine n, a secant modulus, M_1 (see Figure 6.3-32), is drawn to an arbitrary point (P_1,δ_1) on the nonlinear portion of the load-displacement curve. M_1 may be related to the initial modulus, M, by the parameter α_1:

$$M_1 = \alpha_1 M \qquad (6.3\text{-}87)$$

Similarly, a parameter α_c may be obtained from the point of onset of unstable fracture (P_c, δ_c),

$$M_c = \alpha_c M \qquad (6.3\text{-}88)$$

Once α_1 and α_c are determined, n is calculated from

$$n = 1 + \ln\{\alpha_c(1 - \alpha_1)/[\alpha_1(1 - \alpha_c)]\}/\ln(P_1/P_c) \qquad (6.3\text{-}89)$$

The correction factor β is obtained from

$$\beta = [2n/(n + 1)][(1 - \alpha_1)/\alpha_1][P_c/P_1]^{n-1} \qquad (6.3\text{-}90)$$

The major advantage of this technique is that only a single crack length is required to characterize the nonlinear interlaminar fracture of the specimen.

To summarize, three strain energy release rate parameters for experimental characterization of materials exhibiting subcritical crack growth combined with some degree of inelastic material around the crack tip have been defined. These are in increasing order: G_{IISC}, G_{IIC}, and G_{IICNL}. Identical results are obtained for materials exhibiting linear load-displacement response. Typical results are shown in Table 6.3-7 for graphite/epoxy and graphite/PEEK laminates tested over a wide range of cross-head speeds.

Inelastic Material Response. Effects of inelastic material response on mode II interlaminar fracture have been investigated by Chatterjee [80]. A model based on shear deformation plate theory analysis of the ENF specimen includes elastic, perfectly plastic material response that is localized at the crack tip. The model is similar to the Dugdale model [81] developed originally for metals but focuses on shear yielding in the present

application. Within the plastic zone (*PZ*) at the crack tip, the interlaminar shear stress is assumed to be constant and equal to the yield stress, τ_y. The plastic zone is assumed to be narrow in height and centered on the mid-plane. The constraint imposed by the presence of elastic fibers within the composite makes this assumption physically reasonable. Chatterjee [80] derived the following expression for the length (R) of the plastic zone:

$$R = \tau_{BT}a/(\tau_y - \tau_{BT}) = aS/(1 - S) \qquad (6.3\text{-}91)$$

where a is the crack length and τ_{BT} is the nominal shear stress at the mid-plane for a homogeneous ENF beam.

$$\tau_{BT} = 3P/8wh \qquad (6.3\text{-}92)$$

The nondimensional shear stress, S, is defined as

$$S = \tau_{BT}/\tau_y \qquad (6.3\text{-}93)$$

Note that the length of the plastic zone is predicted to be directly proportional to crack length and inversely proportional to yield stress of the material. Based on the crack length dependent plastic zone length in Equation (6.3-91), one would expect an R-curve response in materials undergoing extensive inelastic material response.

The influence of the crack tip on the stress distribution may impose further constraints on the maximum admissible crack length for ENF testing discussed previously. For linear elastic material response, Gillespie et al. [72] have defined the upper bound on crack length based on stress decay of the crack tip singularity. The decay length, λ_{SV}, was found to be proportional to lami-

Table 6.3-7. Rate effects in mode II fracture. AS4/3501-6: $[0]_{24}$; APC-2: $[0]_{26}$. $a/L \approx 0.5$, $L = 51$ mm. Five replicates [40].

Material	Cross-head Rate mm/min	Crack Tip Velocity m/s × 10⁹	G_{IISC} kJ/m²	G_{IIC} kJ/m²	β	G_{IICNL} kJ/m²
AS4/3501-6	0.25	2.63	0.40 ± 0.05	0.53 ± 0.50	0.08 ± 0.02	0.57 ± 0.06
AS4/3501-6	2.5	26.3	0.43 ± 0.08	0.52 ± 0.06	0.06 ± 0.01	0.54 ± 0.05
AS4/3501-6	25	263	0.44 ± 0.07	0.51 ± 0.05	0.05 ± 0.04	0.53 ± 0.04
AS4/3501-6	250	2634	0.45 ± 0.05	0.45 ± 0.05	0	0.45 ± 0.05
APC-2	0.25	2.86	0.96 ± 0.23	1.83 ± 0.27	0.18 ± 0.04	2.16 ± 0.32
APC-2	2.5	28.6	1.00 ± 0.10	1.78 ± 0.15	0.15 ± 0.02	2.05 ± 0.18
APC-2	25	286	0.98 ± 0.15	1.90 ± 0.24	0.12 ± 0.04	2.10 ± 0.27
APC-2	250	2860	1.58 ± 0.15	1.58 ± 0.15	0	1.58 ± 0.15

nate thickness and the ratio of flexural to interlaminar shear modulus and imposes the following constraint on crack length:

Elastic: $a_{max} < L - 1.6\, h[E_1/G_{13}]^{1/2}$ (6.3-94)

Chatterjee [80] proposed a similar argument for the inelastic case:

Inelastic: $a_{max} < L - 5h - aS/(1 - S)$ (6.3-95)

The following expressions for compliance (C_{PZ}) and strain energy release rates (G_{IIPZ}) for ENF specimens with inelastic material response were derived [80]:

$$C_{PZ}/C_{BT} = 1 + 1.5S[1 - S]^{-1} + 0.5S^2[1 - S]^{-2}$$
$$(6.3-96)$$

$$G_{IIPZ}/G_{II}^{BT} = [1 - S]^{-1}   (6.3-97)$$

The beam theory expressions for compliance and strain energy release rate are defined in Equations (6.3-55) and (6.3-57), respectively. Chatterjee [80] also showed that the evaluation of the J-integral evaluated around a contour surrounding the plastic zone gives the same result as Equation (6.3-97).

For brittle thermoset composites, interlaminar fracture occurs at nominal mid-surface shear stress levels far below the yield stress, and the elastic analyses for characterizing mode II interlaminar fracture toughness should be adequate. For tough thermoplastic systems such as graphite/PEEK, the nonlinear response may be significant. For example, when $S = 0.5$, the presence of the plastic zone increases the energy release rate by a factor of two relative to the elastic baseline [see Equation (6.3-97)]. When $S = 0.1$ and 0.2, the compliance of the ENF specimen increases by 2.7 and 6.4 percent from the elastic baseline. The corresponding increase in energy release rates is much greater (11 and 25 percent, respectively). Consequently, according to this analysis, the effect of plastic deformation on energy release rates can be significant even when nonlinearity in load-deformation response recorded experimentally appears to be negligible.

Crack Resistance (R-Curve). Hashemi et al. [47] employed the ELS specimen (see Figure 6.3-19) to investigate mode II interlaminar fracture toughness as a function of crack length. The unique feature of this specimen is that it can be employed for mode I, mode

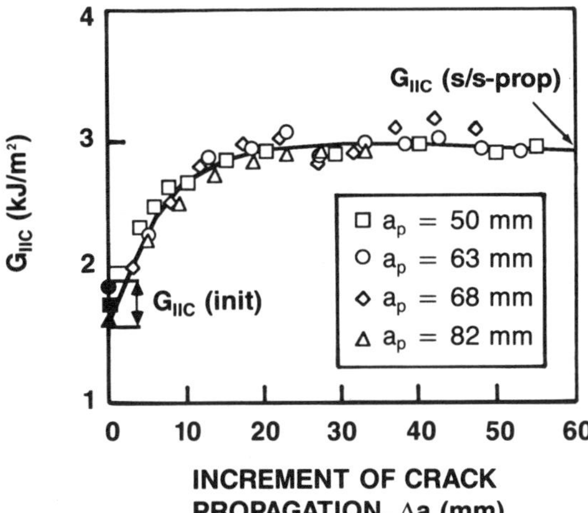

FIGURE 6.3-33. Mode II interlaminar fracture energy of graphite/PEEK exhibits R-curve with increases in crack propagation increment (Δa) during ELS testing [56]. Mode I precrack and cross-head speed = 2 mm/min. Initiation toughness: G_{IIC}(init); steady state/propagation toughness: G_{IIC}(s/s-prop).

II, and mixed mode I/II loadings. In the mode II configuration, stable crack propagation occurs for $a/L > 0.45$. In their experimental study, a mode I precrack was propagated from an aluminum foil insert of length (a_o) to various total crack lengths, a_p, prior to mode II interlaminar fracture testing.

In this regime ($a_p/L > 0.45$), the R-curve is shown in Figure 6.3-33 for graphite/PEEK unidirectional composites. The crack initiates from a relatively low level [G_{IIC}(init) = 1.73 ± 0.04 kJ/m²] and rises rapidly to a steady state propagation level [G_{IIC}(s/s-prop) = 2.89 ± 0.34 kJ/m²]. A unique R-curve is obtained from the various crack lengths investigated when the interlaminar fracture toughness is plotted versus the increment of crack propagation (Δa). The effect of testing at different crack lengths, a_p, is simply to shift the R-curve horizontally. The master curve is shown in Figure 6.3-33.

During mode II interlaminar fracture of graphite/PEEK, the primary mechanism contributing to the increase in fracture energy is the increase in crack tip plasticity that evolves with crack growth. Furthermore, the amount of fiber bridging that occurs during mode I precracking does not appear to significantly change G_{IIC}(init). In contrast, the R-curve observed during mode I interlaminar fracture testing is significantly influenced by the amount of fiber bridging that occurs

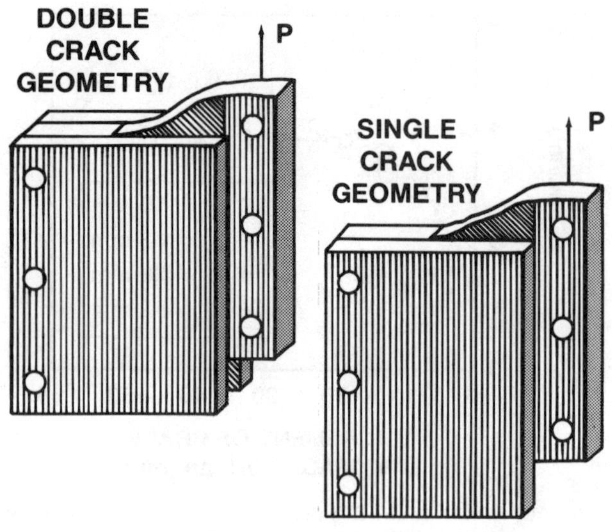

FIGURE 6.3-34. Crack rail shear specimen geometries.

during crack growth. One of the major disadvantages of the ENF specimen is that a complete *R*-curve cannot be obtained, since unstable crack propagation occurs during the test for recommended initial crack lengths. For design purposes, the initiation toughness is very important to characterize. Thin inserts and minimal precracking are recommended for accurate characterization of initiation toughness using the ENF fracture specimen.

6.3.4 Mode III Interlaminar Fracture

In general, a delamination may be subjected to mixed mode I, II, and III crack tip loadings (see Figure 6.3-1). Consequently, test methods for these three characteristic modes of interlaminar fracture are essential to quantify the damage tolerance of the composite structure. In

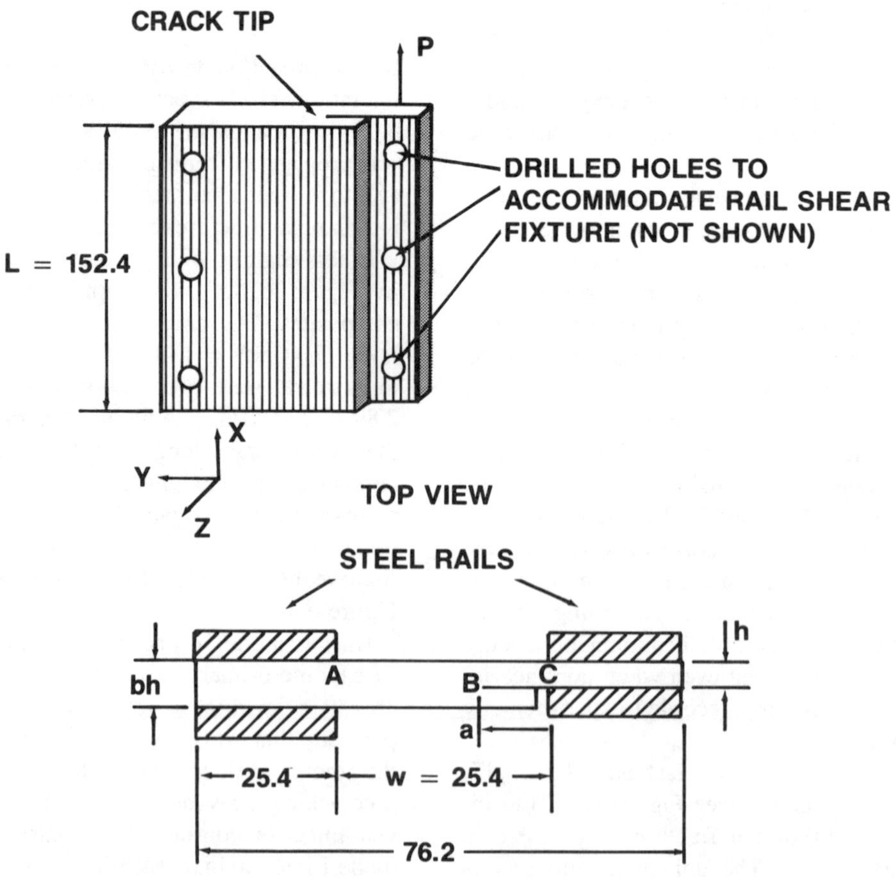

FIGURE 6.3-35. Single crack CRS specimen (all dimensions in mm).

subsections 6.3.2 and 6.3.3, test methods developed to characterize mode I and mode II interlaminar fracture toughness were presented. This subsection shall focus on mode III interlaminar fracture of laminated composites. Mode III crack propagation has been shown to be important in the assessment of defect criticality for delamination between angle plies [82] and for two-dimensional delaminations in bending [83].

The Crack Rail Shear (CRS) specimen has been proposed by Becht and Gillespie [84–86] as a test method to characterize the interlaminar mode III critical strain energy release rate (G_{IIIc}) of continuous fiber reinforced composite materials. As the name implies, the specimen configuration is identical to the ASTM guide D4255-83 [87] (i.e., the two-rail shear test discussed previously) for measuring in-plane shear properties of composite laminates. The specimen utilizes the two-rail shear test fixture and contains embedded Kapton film between designated plies to provide a starter crack for subsequent fracture testing. Figure 6.3-34 shows the specimen geometry. Note that the initial delaminated portions (flaps) are cut away from the inner plies (strap) after processing to accommodate the rail shear fixture. With the rails in place, the specimen is loaded in exactly the same manner as the two-rail shear test.

The following segments will focus on mode III interlaminar fracture characterization using the CRS specimen. A brief summary of the mechanics of the CRS specimen is followed by more detailed information on specimen design considerations and preparation, nonlinear shear response, and the definition of critical load during testing. Test results for the mode III interlaminar fracture characterization of AS4/3501-6 and IM7/8551-7 graphite/epoxy composites are also compared to mode I and mode II toughness data.

Mechanics of the Crack Rail Shear Specimen

This discussion presents the mechanics of the CRS specimen and numerical results used to select appropriate specimen geometries.

Strength of Materials Approach. The double and single crack versions of the CRS specimen are shown in Figure 6.3-34. The specimen geometries are very similar to the two-rail shear test (ASTM guide D4255-83) for measuring the in-plane shear properties of composite laminates [87]. The addition of Kapton film placed between appropriate plies serves as the starter cracks for subsequent fracture testing.

Figure 6.3-35 shows a more detailed cross section of the single crack geometry. For modeling purposes, the specimen is divided into two sections: the undelaminated region, AB, and the delaminated region, BC. An analytical expression for the compliance, C^{SM}, of the CRS specimen derived by Becht [84] is

$$C^{SM} = \frac{1}{Lh}\left[\frac{(w-a)}{bG_{AB}} + \frac{a}{G_{BC}}\right] \qquad (6.3\text{-}98)$$

Here, L is the length of the specimen, h is the thickness of the delaminated section, bh is the total thickness of the undelaminated region, w is the width or distance between rails, a is the crack length, and G_{AB} and G_{BC} are the shear moduli of sections AB and BC, respectively. The superscript SM denotes Strength of Materials, which was utilized to derive Equation (6.3-98). Compliance of the double crack CRS specimen (see Figure 6.3-34) is also defined by Equation (6.3-98) [85].

Implicit in the derivation is the assumption that a pure shear stress state exists within the CRS specimen. To achieve a pure shear stress state, the specimen would have to be of infinite length. This would effectively eliminate flexural deformations and related mode II crack tip loadings. The crack tip singularity and edge effects at $x/L = 0$ and 1 are neglected in this approach. A full derivation of this equation is given in references [84,85].

By utilizing Griffith's energy criterion [Equation (6.3-25)], an expression for the strain energy release rate of the CRS is given as

$$G^{SM} = \frac{P^2(G_{AB}b - G_{BC})}{2nL^2h\,G_{AB}G_{BC}b} \qquad (6.3\text{-}99)$$

Equation (6.3-99) assumes linear elastic response and uniform crack extension across the length, L, of the specimen. In the derivation, the incremental crack surface area, dA, is defined as

$$dA = nLda \qquad (6.3\text{-}100)$$

where da is the incremental change in crack length, $n = 1$ for single crack specimens, and $n = 2$ for double crack specimens. This is the only difference in the strain energy release rate expression [Equation (6.3-99)] for the two CRS geometries shown in Figure 6.3-34.

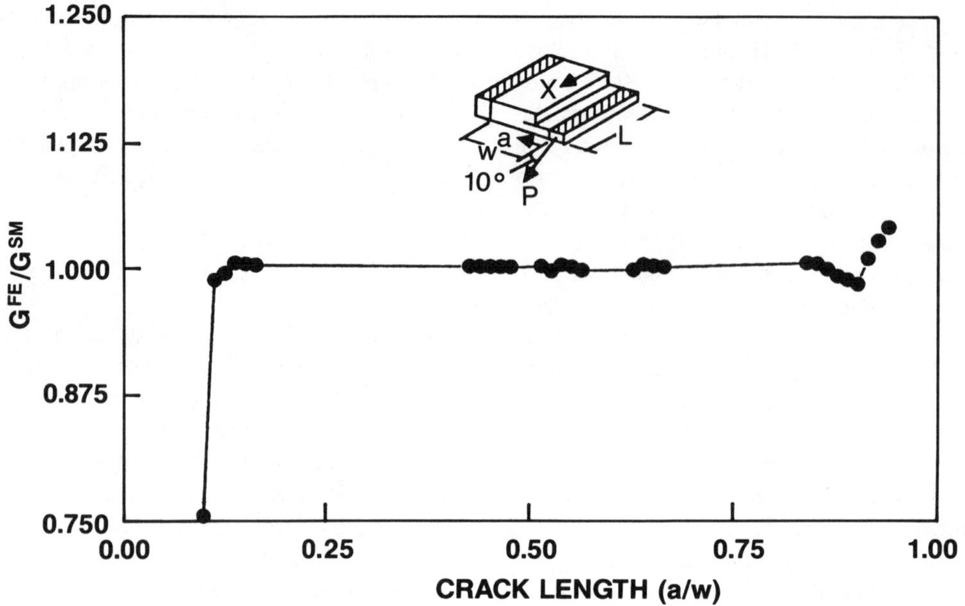

FIGURE 6.3-36. Strain energy release rate results of the quasi-three-dimensional and strength of materials models ($[90°]_{28}$, $b = 2$, double crack geometry).

Crack growth in the CRS specimen is quasi-stable for constant load conditions. For the more common fixed displacement condition, the strain energy release rate can be expressed in terms of the displacement, δ, and specimen compliance:

$$G = \frac{\delta^2}{2nLC^2} \frac{dC}{da} \qquad (6.3\text{-}101)$$

Differentiating Equation (6.3-101) with respect to crack length yields

$$\frac{dG}{da} = \frac{\delta^2}{2nLC^2} \left[\frac{d^2C}{da^2} - \frac{2}{C} \left(\frac{dC}{da} \right)^2 \right] \qquad (6.3\text{-}102)$$

Since the compliance of the CRS specimen is a linear function of crack length [see Equation (6.3-98)], the first term in brackets (i.e., the second derivative of compliance) is always zero. The remaining quantity is always negative, which implies that crack growth under fixed grip conditions is stable. This is in agreement with experimental observations [85,86].

Verification of Strength of Materials Approach. The strength of materials approach yields an expression for the total strain energy release rate but provides no in-formation on the relative magnitudes of the individual components. The purity of the mode III fracture state in the CRS specimen was investigated in references [84,86] using a quasi-three-dimensional linear elastic finite element formulation [88] that discretizes the cross section shown in Figure 6.3-35. Consequently, this analysis also assumes that the specimen is infinitely long and that edge effects (at $x/L = 0$ and 1) are negligible. Based on this model, the CRS specimen was shown to be a pure mode III fracture test. Compliance and strain energy release rate predictions were in good agreement over a large range of crack lengths, with SM predictions based on Equations (6.3-98) and (6.3-99), respectively. This is illustrated in Figure 6.3-36, where the finite element strain energy release, G^{FE}, is normalized by G^{SM} and plotted versus crack length. The two models diverge only when the crack tip is in close proximity to the rails used to introduce load into the specimen. The SM model is an appropriate data reduction scheme for normalized crack lengths greater than 0.15 and less than 0.85 subject to the constraints implicit in the analyses discussed previously. From an experimental viewpoint, crack lengths close to $a/w = 0.5$ are recommended.

A fully three-dimensional linear elastic finite element analysis of the CRS specimen was conducted by Becht and Gillespie [84,86] to investigate the influence of edge effects on the fracture state for the finite length sample. Results based upon a uniform crack extension indicate a small region of mixed mode behavior at traction free edges that decay to a pure mode III fracture state in the interior of the sample. More importantly, the G_{III} distribution along the crack front decreases at the free edges from a maximum plateau region in the interior. The three-dimensional analysis enabled edge effects to be minimized by selecting appropriate specimen lengths. Increasing the length of the specimen [86] minimizes this potential problem while retaining the simplicity of the SM data reduction scheme. The specimen design conducted in reference [86] indicates that a specimen length of 152.4 mm (6.0 in) is sufficient to minimize the influence of the edge effects on mode III fracture characterization.

Experimental Approach

This segment includes discussion on specimen design considerations and preparation, nonlinear shear response, the definition of critical load, and test results. Results for G_{IIIc} of typical graphite/epoxy laminated composites are compared to G_{IC} and G_{IIC}. Post-test interrogation of crack front profiles and fracture surfaces is also presented.

Specimen Design Considerations. The analysis of the CRS specimen is based on linear elastic behavior. In general, composite materials will exhibit nonlinear response when subject to shear loads [89,90]. Since any significant nonlinearity would violate assumptions implicit in the data reduction scheme, minimization of this problem is desirable. In contrast to the unidirectional DCB and ENF specimens, composite laminates containing [±45] sublaminates are required to minimize nonlinear shear response prior to crack propagation. It is also desirable for crack growth to occur in the fiber direction consistent with mode I and mode II characterization. Since the imbedded delamination propagates across the width of the specimen (see Figure 6.3-35), plies adjacent to the delamination plane should have this orientation (defined as 90-degree plies in the subsequent discussion).

The specimen should be designed such that alternate failure mechanisms, such as transverse ply cracking, do not occur prior to crack propagation. The data reduction scheme does not account for this type of damage. The additional energy consumed during testing would overestimate the actual mode III fracture toughness. The thickness of the flap and strap of the CRS as well as the laminate stacking sequence should also be chosen to avoid shear buckling of the specimen prior to crack propagation. A final consideration is laminate anisotropy. The sublaminates of the flap and strap regions, although different in thickness by a factor of b (see Figure 6.3-35), should have identical balanced and symmetric laminate stacking sequences. This eliminates the undesirable coupling phenomenon unique to laminated composites. Furthermore, strain energy associated with residual stresses in laminates would not be released during crack propagation for this class of laminates. The state of residual stress in the flap and strap laminates would be identical before and after delamination growth (i.e., no warpage of the delaminated regions should occur). This would not be the case for hygrothermal loadings, where gradients in moisture content through the thickness of the specimen are present. In this case, warpage of the flap and strap sublaminates would be expected, and the strain energy released from these internal stresses would contribute to the fracture process. It is recommended that gradients in moisture content be minimized by appropriate specimen conditioning. For more detail on CRS specimen design considerations, the interested reader is referred to the original references by Becht and Gillespie [84–86].

In summary, recommended laminate stacking sequences for CRS mode III characterization are of the $[90_n/\pm45_m]_{sp}$ family. During fabrication, a delamination is imbedded between the 90-degree plies in strap and flap sublaminates. Successful specimen designs have been achieved for materials exhibiting G_{IIIc} on the order of 1.1 kJ/m² using $[90_2/\pm45]_{s4}$ stacking sequences. Specifically, the strap laminate is $[90_2/\pm45]_s$ and the flap laminate is $[90_2/\pm45]_{s3}$. In this case, the total thickness of the specimen is four times greater than the strap laminate (i.e., $b = 4$ in Figure 6.3-35). For materials exhibiting greater toughness, alternate laminate constructions may be required.

Specimen Preparation. Two layers of Kapton film should be placed between appropriate plies to provide starter cracks. After machining but prior to end tabbing, individual samples should be clamped and precracked

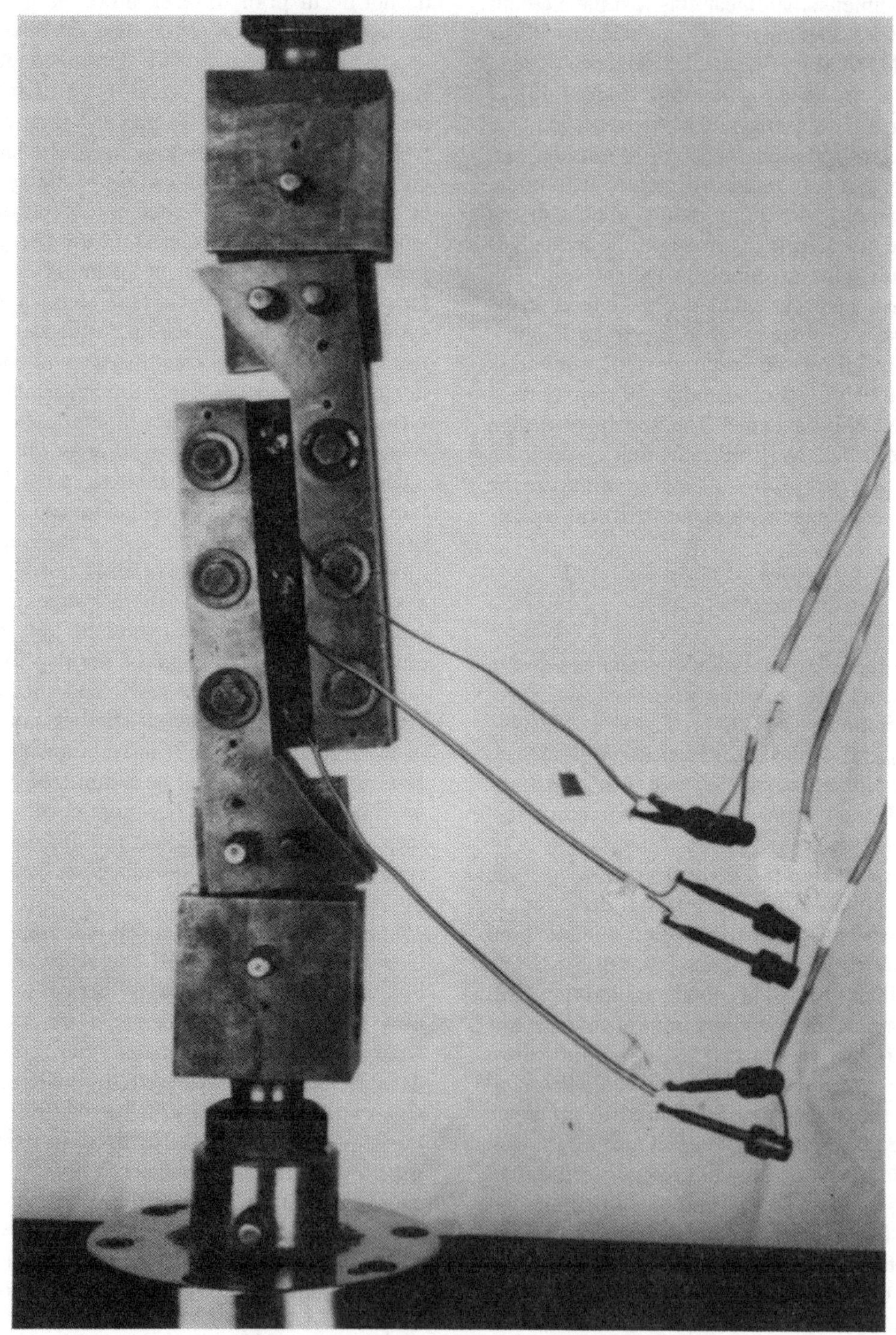

FIGURE 6.3-37. Close-up view of crack rail shear specimen in mounted test frame.

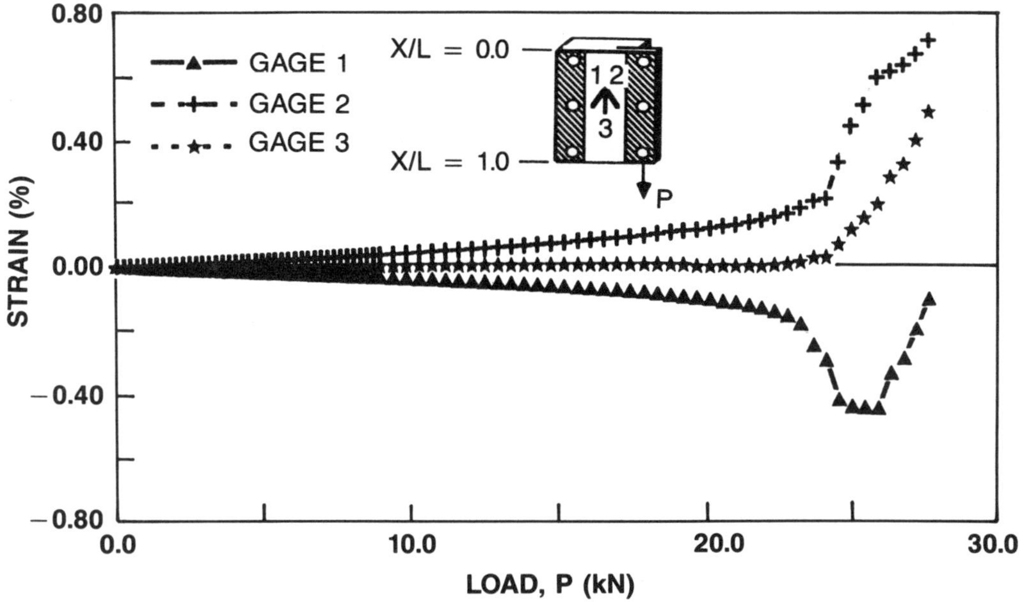

FIGURE 6.3-38. Load-strain response of a 32-ply, AS4/3501-6, $[0_2/\pm45]_{s4}$, $b = 4$, CRS specimen.

across the entire length using a steel wedge or similar device [84,86]. This produces a mode I type precrack uniformly across the length. Precracking is essential to avoid artificially high fracture toughness due to a blunted crack tip commonly embedded in a matrix rich region at the end of the Kapton film [40].

Nondestructive inspection using ultrasonic C-scan techniques (see section 6.2) of the specimens to ensure quality and to determine crack front profiles prior to testing should be routinely performed. This information is very useful for comparison with C-scans of crack front profiles after testing. In addition, accurate placement of strain gages on the specimen surface after precracking is based on the C-scan images. Gages are placed ahead of the crack tip in the uncracked region to detect crack initiation and define critical load values. A typical test specimen mounted in the two-rail shear fixture is shown in Figure 6.3-37.

Critical Load and Shear Modulus Determination. A typical load-strain response of the CRS test is shown in Figure 6.3-38. This figure corresponds to a single crack $[0_2/\pm45]_{s4}$ AS4/3501-6 CRS specimen ($b = 4$). Note that the 0-degree plies are oriented in the length direction of the CRS specimen in this example. Strain instrumentation consists of a three-gage rosette located at the specimen center as shown in Figure 6.3-38. The load-strain response from the $\pm45°$ gages is symmetric

about the load axis, and the $0°$ gage strain is virtually zero throughout most of the loading history. This confirms that a uniform shear stress state does exist and in-plane normal stresses are negligible. The disruption in the curve at approximately 24 kN represents the load at which the delamination front passes under the gages and is defined as the critical load, P_c. For this specimen, gages were placed on the "strap," or continuous side, of the laminate. The strain readings, therefore, increase as the crack passes under the gage due to the reduction in shear area that increases the shear stress in the strap section. Gages placed on the "flap," or discontinuous side of the laminate, should show a zero strain as the crack passes under the gage and load is transferred to the strap.

The specimen should ideally exhibit a linear load-strain response until the crack tip alters the strain field recorded by the gages. When this happens, a sudden increase in strain (by a factor of b) should occur due to the reduction in shear area (by a factor of b). A second region of linear response should then occur. This behavior is depicted in Figure 6.3-39(a) by the solid line. Experimentally, the load-strain response makes a smooth transition between the two linear regions due to the gradual disruption of the strain field by the approaching crack tip singularity. This behavior is illustrated by the dashed line in Figure 6.3-39(a).

A well-defined critical load value is sometimes diffi-

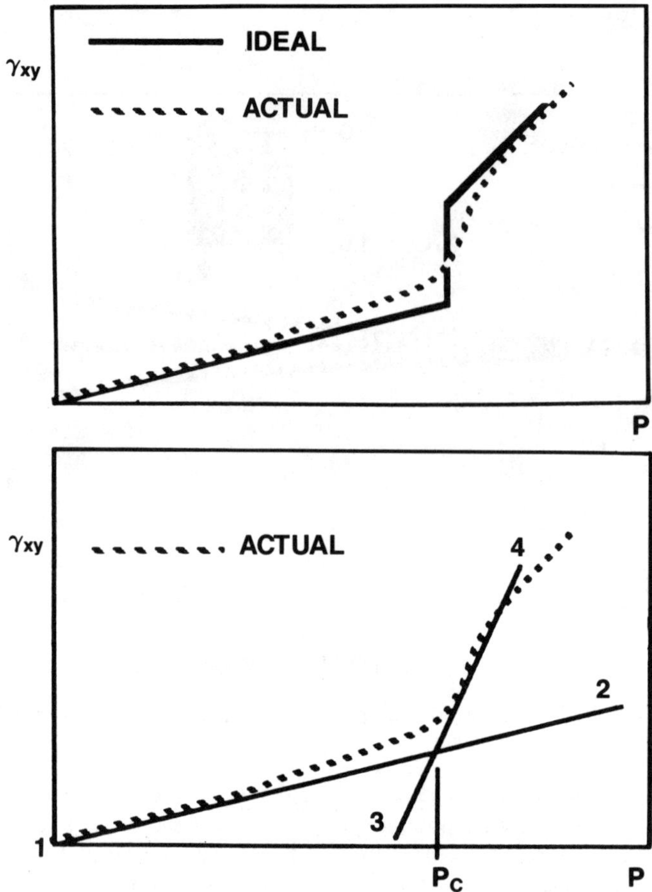

FIGURE 6.3-39. (a) Ideal and actual load-strain response of the CRS specimen and (b) definition of critical load.

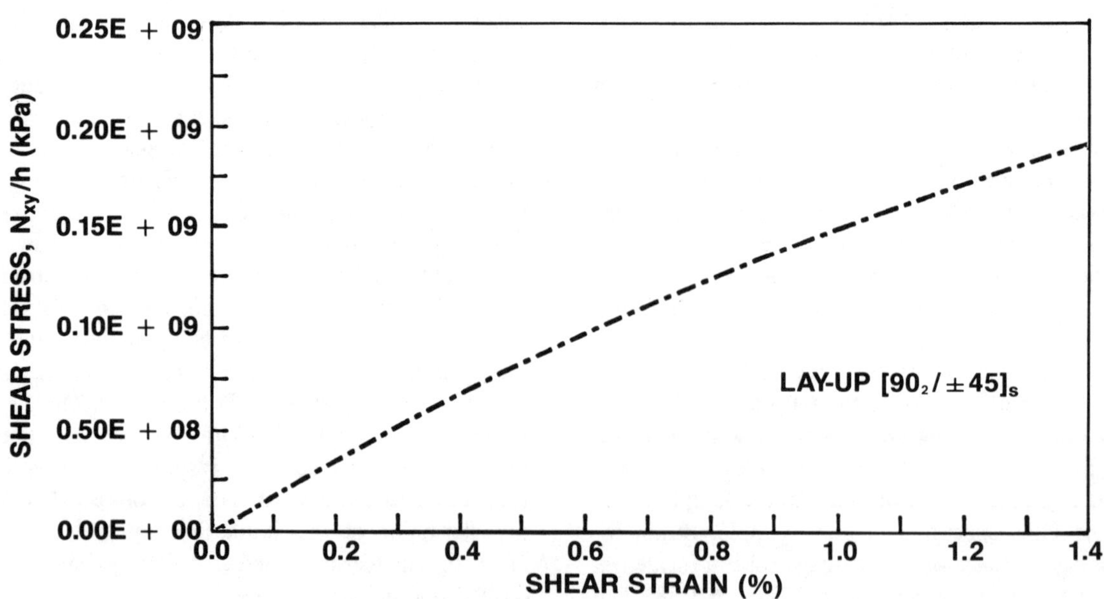

FIGURE 6.3-40. Shear stress-strain response of AS4/3501-6 rail shear test. Lay-up is $[90_2/\pm 45]_s$.

Table 6.3-8. Average G_{IIIc} results for AS4/3501-6 and IM7/8551-7.

Material	Lay-up	Replicates	# of Cracks	P_c (kN)	γ_{AB} (%)	G_{AB} (GPa)	γ_{BC} (%)	G_{BC} (GPa)	G_{IIIc} (J/m²)	CV (%)
AS4/3501-6	$[0_2/\pm45]_{s4}$	3	1	25.9	.229	17.7	1.137	14.3	777	18
	$[90_2/\pm45]_{s4}$	9	1	26.2	.256	17.6	1.306	13.8	922	15
	$[(90/\pm45)_{s3}]_4$	5	1	39.5	.130	22.2	0.597	19.4	616	27
	$[90_2/\pm45]_{s8}$	3	2	54.9	.268	14.0	1.381	13.6	972	17
Overall Average		20							832	24
IM7/8551-7	$[90_2/\pm45]_{s4}$	6	1	37.0	.284	20.4	1.158	18.6	1,095	20

cult to determine because of this response. Hence, the following procedure is recommended to determine critical load in a consistent manner [89]. This is shown in Figure 6.3-39(b). Two lines are drawn on the figure: one tangent to the initial slope of the curve, the other tangent to the curve where the strain gradient is the highest. These lines are labeled 1–2 and 3–4, respectively, in Figure 6.3-39(b). The intersection of these two lines defines the critical load value, P_c, used in Equation (6.3-99) for data reduction purposes.

Also evident in Figure 6.3-38 is a slight nonlinearity in the curve beginning at approximately 14 kN in this sample. The nonlinearity can arise from the approaching crack tip as discussed before or from laminate shear response. To investigate the latter mechanism, uncracked two-rail shear specimens with the same lay-up as the CRS specimens were tested for each material system. Figure 6.3-40 presents the results for an AS4/3501-6, $[90_2/\pm45]_s$ laminate. In this case 90-degree plies are oriented across the width of the specimen (y-direction in Figure 6.3-35). Crack growth is therefore parallel to the fiber direction consistent with DCB and ENF testing. This curve represents an average of three specimens and is a typical response for both material systems investigated. As can be seen, some degree of nonlinearity is present, despite the presence of the $\pm45°$ fibers. The secant shear modulus at 1.0% strain is 80% of the initial modulus. For the IM7/8551-7, the 1.0% secant modulus is approximately 90% of the initial value. The degree of nonlinearity in

Table 6.3-9. Comparison of interlaminar fracture toughness.

	G_{Ic} (J/m²)	G_{IIc} (J/m²)	G_{IIIc} (J/m²)
AS4/3501-6	200	525	832
IM7/8551-7	690	1,100	1,095

the present study is not considered severe. Nevertheless, the data reduction scheme is based on Linear Elastic Fracture Mechanics, and material nonlinearity is accommodated in a consistent manner as discussed later.

An important implication of the nonlinear shear stress-strain response is that the laminate shear modulus is strain dependent. Consequently, the undelaminated region, AB, always has a higher shear modulus than the delaminated region, BC, which is subjected to a higher strain level because of the smaller laminate thickness. To calculate G_{IIIc} for a nonlinear material, the shear moduli in Equation (6.3-99) are defined as the secant moduli that correspond to the strain level at the critical load.

To calculate the secant moduli, the shear stress in each section must first be calculated. For section AB, the effective shear stress, τ_{AB}, is simply

$$\tau_{AB} = P_c/Lbh \qquad (6.3\text{-}103)$$

For section BC the shear stress, τ_{BC}, is

$$\tau_{BC} = P_c/Lh \qquad (6.3\text{-}104)$$

With each of these stress levels known, the secant moduli for each section, G_{AB} and G_{BC}, are then computed as illustrated in Figure 6.3-41 from the corresponding nonlinear shear strains, γ_{AB} and γ_{BC}, respectively. With these quantities known, Equation (6.3-99) is utilized to determine G_{IIIc}.

Mode III Critical Strain Energy Release Rate Results. Results used for mode III characterization of AS4/3501-6 and IM7/8551-7 are summarized in Table 6.3-8. Results include the lay-up, critical load, shear strain, and secant moduli at the critical load, G_{IIIc}, and coefficient of variation (CV). The overall average G_{IIIc} of AS4/3501-6 is based upon 20 tests and is equal to 832

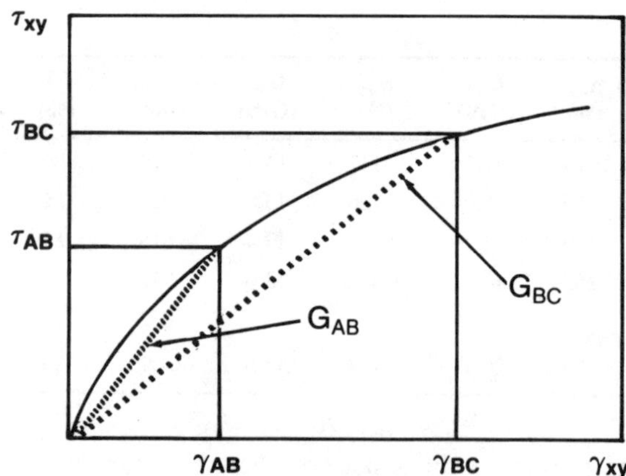

FIGURE 6.3-41. Determination of secant shear moduli from nonlinear response of CRS specimens.

J/m² with a *CV* of 24%. The average G_{IIIC} of IM7/8551-7 is based upon six tests and is equal to 1,095 J/m² (*CV* of 20%). Table 6.3-9 gives G_{IC} and G_{IIC} data for these materials for comparison [40,86,91]. G_{IIIC} for AS4/3501-6 is on the average 1.6 times G_{IIC} and 4.2 times G_{IC}. For the tougher IM7/8551-7, G_{IIIC} is approximately equal to G_{IIC} and is 1.6 times greater than G_{IC}.

It should be noted that delaminations did not stay in the original plane in any of the AS4/3501-6 specimens but branched to the adjacent 45° layer interface. This phenomenon was quite unexpected for the $[90/\pm45]_s$ type laminates, since it was believed that the 90° fibers (fiber direction parallel to crack growth direction) would constrain the delamination to the original plane. Delamination branching was fully expected for the $[0_2/\pm45]_{s4}$ lay-up, since the crack front's path to adjacent planes was not constrained by the fibers. Multiple delaminations were also observed in some specimens. This behavior was previously reported [83] for disbonds in laminates subjected to transverse loading. Evidently, a higher strain energy release rate was available for crack formation at these interfaces, perhaps due to the thermal residual stresses that were induced upon consolidation. If this were the case, G_{IIIC} values reported herein would be lower than the actual mode III toughness value, since thermal effects are not included in the data reduction [92–94].

In contrast to the AS4/3501-6 specimens, the IM7/8551-7 CRS specimens exhibited self-similar crack growth; i.e., no crack branching or deviation from the original plane was observed. This behavior is attributed to the higher toughness of the matrix.

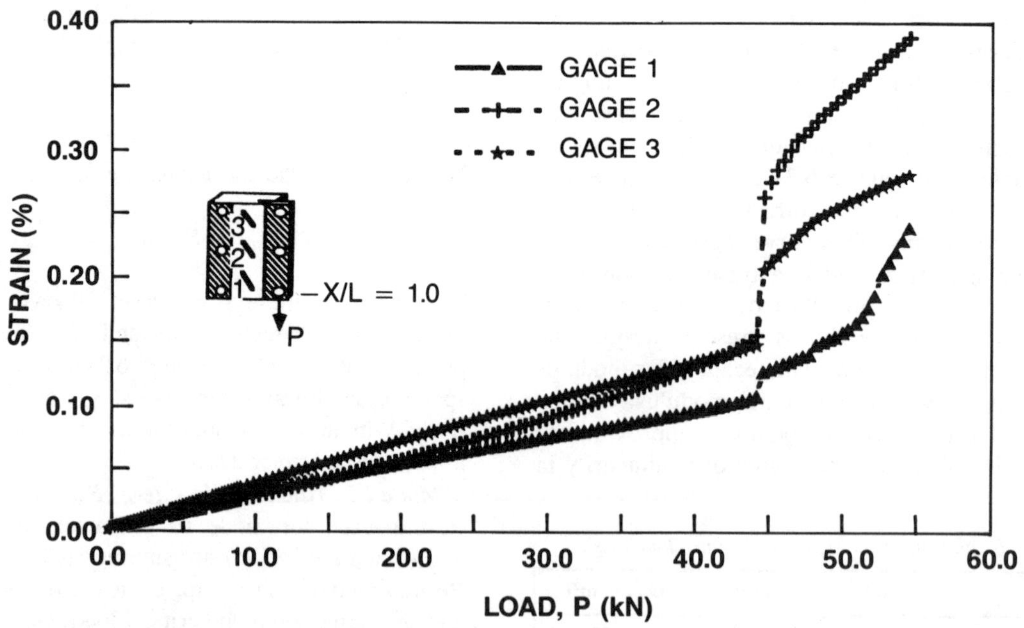

FIGURE 6.3-42. Load-strain response of CRS specimen with strain gages positioned along the length at x/L = 0.2, 0.5, 0.8.

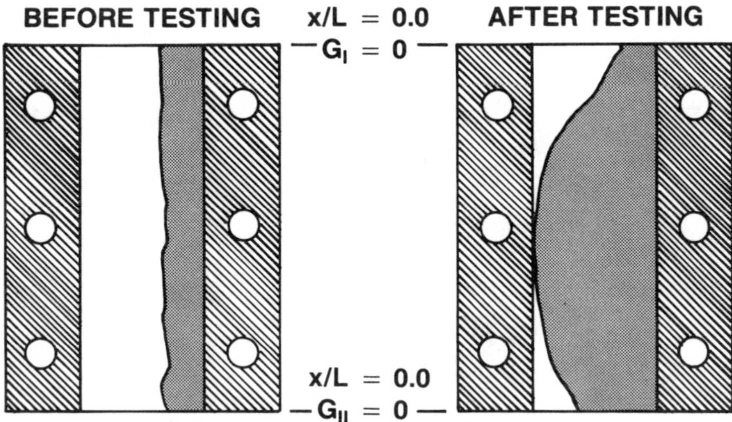

FIGURE 6.3-43. CRS specimen crack front profile before and after testing.

Results from the AS4/3501-6 double crack specimens are in general agreement with the thinner single crack specimens. These specimens, however, did not delaminate symmetrically; i.e., crack growth began on one side first and was followed by the second crack. Double crack CRS specimens also required very high loads for delamination (see Table 6.3-8), which caused some degree of slippage between the rails and end tabs prior to failure. For these reasons, the single crack CRS specimen geometry is recommended for fracture testing.

Crack Front Profile Measurements. To gain some insight into the crack front profile during the fracture process, strain gages were positioned along the length of some specimens as shown in Figure 6.3-42. Single strain gages oriented 45° to the load axis were located at $x/L = 0.2, 0.5,$ and 0.8. Each gage yielded a very similar load-strain response, indicating a uniform shear strain across the length of the sample. Furthermore, all three gages indicate delamination growth at virtually the same load, implying uniform crack growth. Ultrasonic inspection of the specimen after testing, however, revealed a crack front profile shown schematically in Figure 6.3-43. Preferential growth in the center of the specimen at $x/L = 0.5$ was observed for every specimen tested. Interestingly, the free edge at $x/L = 0.0$ showed very little crack growth, while the region at $x/L = 1.0$ exhibited an intermediate increase in crack length. This experimental result is consistent with the numerical predictions of the strain energy release rate profile [84,86], where a higher strain energy release rate in the interior region initiates delamination growth.

From the physics of the problem, it can be seen that growth at the traction free edges will not begin until the delamination has grown enough in the center to critically load these edge regions. Observations of the crack front profiles after testing indicate that the width of the CRS specimen is not sufficient to allow this process to fully develop along the length. Hence, only a small amount of crack growth is recorded at the free edges. G_{IIIc} measurements based on visual crack growth at these edges may thus greatly overestimate the critical load and fracture toughness. Consequently, strain gages are recommended for defining critical loads in this test method.

Fractographic Analysis. Upon completion of the experimental test program, representative samples from each material system were subjected to a fractographic analysis. This entailed the removal of small areas of the test specimens and observation of the fracture surface morphology under a Scanning Electron Microscope (SEM). Figure 6.3-44(a) shows a 640× magnification of a $[90_2/\pm 45]_{s4}$, AS4/3501-6 CRS specimen at $x/L = 0.5$. Crack propagation is from left to right at an angle of approximately 10°. An obvious conclusion is that delamination is not between 90° plies. Instead, the crack has branched across these plies and propagated along the 90°/45° interface. Figure 6.3-44(b) shows an AS4/3501-6 mode II fracture surface for comparison [36]. Hackle formations in the matrix characteristic of a shear failure are evident in both materials. This fracture behavior is characteristic of a brittle material system.

Figure 6.3-45(a) shows an SEM photo of an IM7/8551-7 CRS specimen. This picture is a 1,250× magnification of an area near the center, $x/L = 0.5$. The

(a)

(b)

FIGURE 6.3-44. (a) Mode III fracture surface of a 32-ply, $[90_2/\pm 45]_{s4}$, AS4/3501-6 CRS specimen. Photo is taken of an area located at $x/L = 0.5$ (640× magnification). (b) Mode II fracture surface of AS4/3501-6 ENF specimen [36] (640× magnification).

(a)

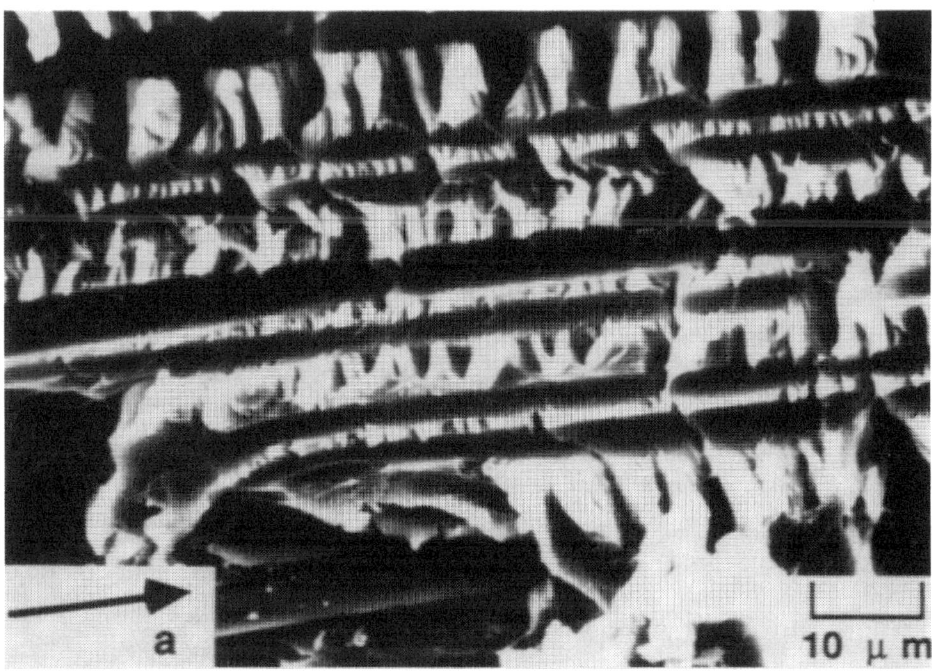

(b)

FIGURE 6.3-45. (a) Mode III fracture surface of a 32-ply, $[90_2/\pm45]_{s4}$, IM7/8551-7 CRS specimen. Photo is taken of an area located at $x/L = 0.5$ (1250× magnification). (b) Mode II fracture surface of IM7/8551-7 ENF specimen (1250× magnification).

fracture behavior of this material system is quite different from AS4/3501-6. The hackles seem to be drawn out more, showing a higher degree of matrix deformation. This is not surprising, since the 8551-7 is a toughened matrix. An ENF fracture surface is shown for comparison in Figure 6.3-45(b). The leading edges of the hackles in both mode II and mode III are basically orthogonal to the direction of crack growth.

6.3.5 References

1. WILKINS, D. J., J. R. Eisenmann, R. A. Camin, W. S. Margolis, and R. A. Benson. "Characterizing Delamination Growth in Graphite-Epoxy," *Damage in Composite Materials*, ASTM STP 775, p. 168 (1980).

2. TRETHEWEY, B. R., J. W. Gillespie, Jr., and L. A. Carlsson. "Mode II Cyclic Delamination Growth," *J. Composite Materials*, 22:459 (1988).

3. MARTIN, R. H. and G. B. Murri. "Characterization of Mode I and Mode II Delamination Growth and Thresholds in Graphite/PEEK Composites," *Composite Materials: Testing and Design*, ASTM STP 1059 (1990).

4. WHITCOMB, J. D. "Finite Element Analysis of Instability Related Delamination Growth," *J. Composite Materials*, 15:403 (1981).

5. ROTHSCHILDS, R. J., J. W. Gillespie, Jr., and L. A. Carlsson. "Instability Related Delamination Growth in Thermoset and Thermoplastic Composites," *Composite Materials: Testing and Design (Eight Conference)*, ASTM STP 972, p. 161 (1988).

6. GILLESPIE, J. W., Jr. and L. A. Carlsson. "Buckling and Growth of Delaminations in Thermoset and Thermoplastic Composites," *J. Eng. Mater. Tech.*, in press (1990).

7. BROCK, D. "Elementary Engineering Fracture Mechanics," Martinus Nijhoff (1982).

8. GRIFFITH, A. A. "The Phenomena of Rupture and Flow in Solids," *Phil. Trans. Roy. Soc. of London*, A(221):163 (1921).

9. RICE, J. R. "A Path Independent Integral and the Approximative Analysis of Strain Concentrations by Notches and Cracks," *J. Appl. Mech.*, 35:379 (1968).

10. WESTERGAARD, H. M. "Bearing Pressures and Cracks," *J. Appl. Mech.*, 61:A49 (1939).

11. SIH, G. C., P. C. Paris, and G. R. Irwin. *Int. J. Fracture Mech.*, 1(3):189 (1965).

12. LEKHNITSKII, S. G. *Theory of Elasticity of an Anisotropic Body*. MIR Publishers, Moscow (1981).

13. JONES, R. M. *Mechanics of Composite Materials*. McGraw-Hill (1975).

14. SIH, G. C. and H. Liebowitz. "Mathematical Theories of Brittle Fracture," in *Fracture, Vol. 2*. H. Liebowitz, ed., Academic Press, New York (1968).

15. IRWIN, G. R. "Fracture," *Handbuch der Physik, Vol. 6, Flügge edition*. Springer, Berlin, Germany, p. 551 (1958).

16. RYBICKI, E. F. and M. F. Kanninen. "A Finite Element Calculation of Stress Intensity Factors by a Modified Crack Closure Integral," *Eng. Fracture Mech.*, 9:931 (1977).

17. WANG, A. S. D. and F. W. Crossman. "Initiation and Growth of Transverse Cracks and Edge Delamination in Composite Laminates. Part I. An Energy Method," *J. Composite Materials*, Supplement, 14:71 (1980).

18. WILLIAMS, J. G. "On the Calculation of Energy Release Rates for Cracked Laminates," *Int. J. Fracture*, 36:101 (1988).

19. SOU, Z. "Delamination Specimens for Orthotropic Materials," Report MECH-135, Division of Applied Sciences, Harvard University, Cambridge, MA (November 1988).

20. HUTCHINSON, J. W. Personal communication.

21. WILLIAMS, M. L. "The Stress around a Fault or Crack in Dissimilar Media," *Bulletin of the Seismological Society of America*, 49:199–204 (1959).

22. ENGLAND, A. H. "A Crack between Dissimilar Media," *Journal of Applied Mechanics*, 32:400–402 (1965).

23. ERDOGAN, F. "Stress Distribution in Bounded Dissimilar Materials with Cracks," *J. Applied Mechanics*, 32:403–410 (1965).

24. ERDOGAN, F. and G. D. Gupta. "Layered Composites with an Interface Flaw," *International Journal of Solids and Structures*, 7:1089–1107 (1971).

25. SUN, C. T. and M. G. Manoharan. "Strain Energy Release Rates of an Interfacial Crack Between Two Orthotropic Solids," *Proceedings of the American Society for Composites, 2nd Technical Conference, September 23–25, 1987*, Technomic Publishing Co., Inc., pp. 49–57 (1987).

26. RAJU, I. S., J. H. Crews, Jr., and M. A. Aminpour. "Convergence of Strain Energy Release Rate Components for Edge-Delaminated Composite Laminates," *Engineering Fracture Mechanics*, 30:383–396 (1988).

27. PIPES, R. B. and H. J. Pagano. "Interlaminar Stresses in Composite Laminates under Uniform Axial Extension," *J. Composite Materials*, 4:538 (1970).

28. BASCOM, W. D., R. J. Bitner, R. J. Moulton, and A. R. Siebert. "The Interlaminar Fracture of Organic Matrix Woven Reinforced Composites," *Composites*, 11:9 (1980).

29. Minutes of Meeting Between ASTM Task Group D30.02.02, European Group of Fracture and Japan Industrial Standards Group for Coordination of Standards Being Written for Interlaminar Fracture Toughness, Orlando, FL (November 1989).

30. WHITNEY, J. M. "Stress Analysis of the Double Cantilever Beam Specimen," *Composite Science and Technology*, 23:201–219 (1985).

31. CREWS, J. H., JR., J. N. Shivakumar, and I. S. Raju. "Factors Influencing Elastic Stresses in Double Cantilever Beam Specimens," NASA TM-89033 (November 1986).

32. RAJU, I. S., K. N. Shivakumar, and J. H. Crews, Jr. "Three-Dimensional Elastic Analysis of a Composite Double Cantilever Beam Specimen," *AIAA/ASME/ASCE/AHS 28th Struc-*

tures, Structural Dynamics, and Materials Conference, Monterey, CA, AIAA Paper 87-0864 (April 6–8, 1987).

33. CREWS, J. H., JR., K. N. Shivakumar, and I. S. Raju. "A Fiber Resin Micromechanics Analysis of the Delamination Front in a DCB Specimen," NASA TM 100540 (January 1988).

34. BERRY, J. P. "Determination of Fracture Surface Energies by the Cleavage Technique," *J. Appl. Phys.*, 34:62 (1963).

35. DAVIES, P., and M. L. Benzeggagh. "Interlaminar Mode I Fracture Testing," in *Application of Fracture Mechanics to Composite Materials*, K. Friedrich, ed., Elsevier Science, p. 81 (1989).

36. ASTM Meeting Minutes D30.02.02 Task Group on Interlaminar Fracture Toughness, Nugget Hotel, Sparks, NV (April 26, 1988).

37. "A Protocol for Interlaminar Fracture Testing," *2nd Revised ed.* European Group on Fracture (1990).

38. DEVITT, D. F., R. A. Shapery, and W. L. Bradley. "A Method for Determining the Mode I Delamination Fracture Toughness of Elastic and Viscoelastic Composite Materials," *J. Composite Materials*, 14:270 (1980).

39. WILSON, D. W. and L. A. Carlsson. "Mechanical Characterization of Composite Materials," in *Physical Methods of Chemistry*, B. W. Rossiter, J. F. Hamilton, and R. C. Baetzold, eds., John Wiley & Sons, New York, in press.

40. CARLSSON, L. A., J. W. Gillespie, Jr., and B. R. Trethewey. "Mode II Interlaminar Fracture of Graphite/Epoxy and Graphite/PEEK," *J. Reinf. Plast. Compo.*, 5:170 (1986).

41. "Test Procedures for the Double Cantilever Beam (DCB) Test," ASTM D30.02 Round Robin, ASTM (1987).

42. CARLSSON, L. A. and R. B. Pipes. *Experimental Characterization of Advanced Composite Materials*. Prentice-Hall (1987).

43. WHITNEY, J. M., C. E. Browning, and W. Hoogsteden. "A Double Cantilever Beam Test for Characterizing Mode I Delamination of Composite Materials," *J. Reinf. Plast. Compo.*, 1:297 (1982).

44. JOHNSON, W. S. and P. D. Mangalgiri. "Investigation of Fiber Bridging in Double Cantilever Beam Specimens," NASA TM 87716 (April 1986).

45. CRICK, R. A., D. C. Leach, P. J. Meakin, and D. R. Moore. "Interlaminar Fracture Morphology of Carbon Fiber/PEEK Composites," *J. Materials Science*, 22:2094 (1987).

46. FRIEDRICH, K., L. A. Carlsson, A. J. Smiley, R. Walter, and J. W. Gillespie, Jr. "Mechanisms for Rate Effects on Interlaminar Fracture Toughness of Carbon/Epoxy and Carbon/PEEK Composites," *J. Materials Science*, 24: 3387–3398 (1989).

47. HASHEMI, S., A. J. Kinloch, and J. G. Williams. "The Effects of Geometry, Rate and Temperature on the Mode I, Mode II and Mixed-Mode I/II Interlaminar Fracture of Carbon-Fiber/Poly(ether-etherketone) Composites," *J. Composite Materials*, in press (1990).

48. Boeing Specification Support Standard BBS 7260.

49. "Standard Tests for Toughened Resin Composites," NASA RP 1092 (1982).

50. EVANS, R. E. and J. E. Masters. "A New Generation of Epoxy Composites for Primary Structural Applications: Materials and Mechanics," ASTM STP 937, p. 413 (1987).

51. BERG, C. A., J. Tirosh and M. Israeli. "Analysis of Short Beam Bending of Fiber Reinforced Composites," ASTM STP 497, p. 206 (1972).

52. WHITNEY, J. M. and C. E. Browning. "On Short Beam Shear Tests for Composite Materials," *Exp. Mech.*, 3:294 (1985).

53. WHITNEY, J. M. "Elasticity Analysis of Orthotropic Beams Under Concentrated Loads," *Comp. Sci. & Tech.*, 22:167 (1985).

54. BARRETT, J. D. and R. O. Foschi. "Mode II Stress Intensity Factors for Cracked Wood Beams," *Eng. Fracture Mech.*, 9:371 (1977).

55. RUSSELL, A. J. and K. N. Street. "Factors Affecting the Interlaminar Fracture Energy of Graphite/Epoxy Laminates," *Progress in Science and Engineering of Composites*, T. Hayashi, K. Kawata and S. Umekawa, eds., ICCM-IV, Tokyo, p. 279 (1982).

56. BRADLEY, W. L. Presentation at ASTM Task Group Meeting, Committee D-30, Charleston, SC (April 30, 1986).

57. PREL, Y. J., P. Davies, M. L. Benzeggagh, and F. X. de Charentenay. "Mode I and Mode II Delamination of Thermosetting and Thermoplastic Composites," paper presented at an *ASTM Symposium on Composite Materials Fatigue and Fracture*, Cincinnati, OH (April 27–28, 1987).

58. BENZEGGAGH, M. L., Y. J. Prel, and F. X. de Charentenay. "Instrumentation of Mode I and Mode II Tests for Crack Tip Strain Profile Study Under Static and Cyclic Loading," *Proceedings of the First European Conference on Composite Materials*, p. 291 (September 1985).

59. VU-KHANH, T. "Crack-Arrest Study in Mode II Delamination in Composites," *Poly. Comp.*, 8:331 (1987).

60. MAIKUMA, H., J. W. Gillespie, Jr., and J. M. Whitney. "Analysis and Experimental Characterization of the Center Notch Flexural Test Specimen for Mode II Interlaminar Fracture," *J. Composite Materials*, 23(8):756–786 (1989).

61. MAIKUMA, H., J. W. Gillespie, Jr., and D. J. Wilkins. "Mode II Interlaminar Fracture of the Center Notch Flexural Specimen Under Impact Loading," *J. Composite Materials*, 24(2):124–149 (1990).

62. CREWS, J. H. and J. R. Reeder. "A Mixed-Mode Bending Apparatus for Delamination Testing," NASA TM 100662 (August 1988).

63. CARLSSON, L. A. and J. W. Gillespie, Jr. "Mode II Interlaminar Fracture," *Application of Fracture Mechanics to Composite Materials*, Volume Editor: K. Friedrich, Elsevier Book Series on Composite Materials; Series Editor: R. B. Pipes, Elsevier Science Publishers, Amsterdam (1989).

64. FRIEDRICH, K., L. A. Carlsson, J. W. Gillespie, Jr., and J. Karger-Kocsis, "Fracture of Thermoplastic Composites," *Thermoplastic Composite Materials*, Volume Editor: L. A.

Carlsson, Elsevier Book Series on Composite Materials; Series Editor: R. B. Pipes, Elsevier Science Publishers, Amsterdam (1990).

65. O'BRIEN, T. K., G. B. Murri, and S. A. Salpekar. "Interlaminar Shear Fracture Toughness and Fatigue Thresholds for Composite Materials," NASA TM 89157 (1987).

66. CARLSSON, L. A., J. W. Gillespie, Jr., and R. B. Pipes. "On the Analysis and Design of the End Notched Flexure (ENF) Specimen for Mode II Testing," *J. Composite Materials*, 20:594 (1986).

67. TIMOSHENKO, S. P. and J. N. Goodier. *Theory of Elasticity, 3rd ed*. McGraw-Hill (1970).

68. TIMOSHENKO, S. P. *Strength of Materials, Part I*. Krieger Publishing (1984).

69. GILLESPIE, J. W., JR, L. A. Carlsson, and R. B. Pipes. "Finite Element Analysis of the End Notched Flexure (ENF) Specimen for Measuring Mode II Fracture Toughness," *Comp. Sci. & Tech.*, 26:177 (1986).

70. GREENBERG, M. D. *Foundations of Applied Mathematics*. Prentice-Hall (1978).

71. RUSSELL, A. J. and K. N. Street. "Moisture and Temperature Effects on the Mixed-Mode Delamination Fracture of Unidirectional Graphite/Epoxy Delamination and Debonding of Materials," ASTM STP 876, p. 349 (1985).

72. GILLESPIE, J. W., JR., L. A. Carlsson, and B. R. Trethewey. "Finite Element and Plate Theory Based Design and Data Reduction of the ENF Fracture Specimen," *Proceedings from the 2nd Conference on Composite Materials*, sponsored by the American Society for Composites and the University of Delaware, Technomic Publishing Co., Inc., Lancaster, PA, p. 399 (1987).

73. CARLSSON, L. A., J. W. Gillespie, Jr., and J. M. Whitney. "Numerical and Analytical Evaluation of the ENF Fracture Specimen," paper presented at *ASME Winter Annual Meeting*, Anaheim, CA (December 7–12, 1986).

74. SALPEKAR, S. A., I. S. Raju, and T. K. O'Brien. "Strain Energy Release Rate Analysis of End-Notched Flexure Specimen Using the Finite Element Method," NASA TM-100494 (1987).

75. MURRI, G. B. and T. K. O'Brien. "Interlaminar G_{IIc} Evaluation of Toughened Resin Composites Using the End-Notched Flexure Test," AIAA-85-0647, *Proceedings of the 26th AIAA/ASME/ASCE/ASH Conference on Structures, Structural Dynamics and Materials*, Orlando, FL, p. 197 (April 1985).

76. ASTM Standard E 399–83 (1987).

77. Liebowitz, H. and J. Eftis. "On Nonlinear Effects in Fracture Mechanics," *Eng. Fracture Mechanics*, 3:267 (1971).

78. EFTIS, J., D. L. Jones, and H. Liebowitz. "On Fracture Toughness in the Nonlinear Range," *Eng. Fracture Mechanics*, 7:491 (1975).

79. WESTERLIND, B. S., L. A. Carlsson, and Y. M. Anderson. "Fracture Toughness of Liner Board Evaluated by the J-Integral," submitted to *J. Materials Sci.* (1990).

80. CHATTERJEE, S. N. "Analysis of Test Specimens for Mode II Fracture Toughness," *J. Composite Materials*, in press (1990).

81. DUGDALE, D. S. "Yielding of Steel Sheet Containing Slits," *J. Mechanics and Physics of Solids*, 8:100–104 (1960).

82. WANG, S. S. "Fracture Mechanics for Delamination Problems in Composite Materials," *J. Composite Materials*, 17:210 (1983).

83. CHATTERJEE, S. N., W. A. Dick, and R. B. Pipes. "Mixed-Mode Delamination Fracture in Laminated Composites," *Comp. Sci. & Tech.*, 25:49 (1986).

84. BECHT, G. J. "An Investigation of Interlaminar Fracture under Mode III Loading," Master's Thesis, University of Delaware (1988).

85. BECHT, G. J. and J. W. Gillespie, Jr. "Design and Analysis of the Crack Rail Shear Specimen for Mode III Interlaminar Fracture," *Comp. Sci. Tech.*, 31:143 (1988).

86. BECHT, G. J. and J. W. Gillespie, Jr. "Numerical and Experimental Evaluation of the Mode III Interlaminar Fracture Toughness of Composite Materials," *Polymer Composites*, 10(5) (1989).

87. *ASTM Standards and Literature References for Composite Materials, First Edition*, ASTM, Philadelphia, PA (1987).

88. MOELLER, P. W. "CCM-Q3D Users Manual," University of Delaware Center for Composite Materials Internal Report, Newark, DE (1987).

89. WHITNEY, J. M., I. M. Daniel, and R. B. Pipes. *Experimental Mechanics of Fiber Reinforced Composite Materials*, Society for Exp. Mech. Monograph No. 4, (revised edition), Prentice-Hall, Englewood Cliffs, NJ (1984).

90. BOGETTI, T. A. "Predicting Nonlinear Behavior in Laminated Composites," Masters Thesis, Department of Mechanical Engineering, University of Delaware (1986).

91. GILLESPIE, J. W., JR., L. A. Carlsson, and A. J. Smiley. *Comp. Sci. Tech.*, 28:49 (1987).

92. CHAPMAN, T. J. "The Effects of Cooling Rate on Residual Stresses and Mode I Fracture Toughness of Thermoplastic Composite Materials," Masters Thesis, Department of Mechanical Engineering, University of Delaware (1988).

93. O'BRIEN, T. K., I. S. Raju, and D. P. Garber. NASA TM 86437, NASA (1985).

94. WHITNEY, J. M. *Sixth International Conference on Composite Materials: Second European Conference on Composite Materials*, Elsevier Applied Science, Vol. 3 (1987).